IDEAS THAT CHANGED THE WORLD

THE INDUSTRIAL REVOLUTION

ILLUSTRATED BY ROBERT INGPEN
TEXT BY
PHILIP WILKINSON & MICHAEL POLLARD

CHELSEA HOUSE PUBLISHERS
New York • Philadelphia

First published in the United States in 1995 by
Chelsea House Publishers

First Printing
1 3 5 7 9 8 6 4 2

Simplified text and captions by **Michael Pollard**
based on the *Encyclopedia of Ideas that Changed the
World* by Robert Ingpen & Philip Wilkinson

Editor	Diana Briscoe
Project Editor	Claire Watts
Designer	Design 23
Art Director	John Strange
Design Assistants	Karen Fergusom
	Victoria Furbisher
DTP Manager	Keith Bambury
Editorial Director	Pippa Rubinstein

ISBN 0–7910–2767–8

Printed in Italy

CONTENTS

◆

Introduction

Which are the really great inventions? Perhaps some of the greatest are the ones that make practical use of basic scientific principles – the physical facts that make our world the way it is.

Thousands of years ago, for example, people discovered the basics of mechanics. This led to the invention of a number of simple machines – levers, wedges, and pulleys – that made lifting and moving heavy loads many times easier than it had been before. All kinds of tasks, especially ones to do with building, were made more straightforward. The ancient Egyptians would have found it difficult to build their pyramids without levers and wedges; the cathedral-builders of the Middle Ages found pulleys vital when constructing their towers. We still use these tools today.

There was a similar leap forward in the eighteenth and nineteenth centuries when scientists like Alessandro Volta and Michael Faraday were experimenting with a new source of power: electricity. It was not long before their experiments found their way out of the laboratory into the factory and the home.

Faraday worked out how to generate an electric current using magnetism in the 1820s. Throughout the nineteenth century many other scientists and inventors followed up his work, creating a series of devices, from generators to light bulbs, which made electricity the important source of power it is today. The dim, gas-lit streets of cities were bathed in electric light; more and more homes were connected to the electricity supply; factories no longer relied on temperamental steam engines and went over to the power source that could be turned on with the touch of a button.

In our own century the link between science and power has been

clearest in the discoveries of the nuclear physicists. By working out the structure of the atom – the basic building block of matter – and how to split it, they unleashed a form of energy with a power undreamed of before.

As the physicist Albert Einstein realized, this power could be used for evil as well as for good, creating bombs more destructive than anything that went before. Even nuclear power stations have to be operated according to strict rules if they are to remain safe.

The work of scientists has led to some of the most influential human inventions. In farming, medicine, power generation and the production of plastics, scientists have led the way, and inventors have followed with new products. Together they have transformed the world.

PHILIP WILKINSON

SCIENTIFIC FARMING

Between 1600 and 1900, the population of the world doubled, and farmers had to face the challenge of feeding and clothing billions more people. To do this, they had to update their working methods, improve their crops, and increase their yields.

I n 1600, farming all over the world was carried on in much the same way as it had been for thousands of years. The land was divided into small areas, each cultivated by a farmer and his family. In Europe, these areas were often strips just large enough to provide food for the family's own needs. Each farming family also kept a few animals such as sheep or pigs on "common" land that was set aside for use by everyone.

Life on these small farms was hard. Oxen pulled the plows, but almost all the other work was done by the farmer and his family. They sowed seed by "broadcasting" it, scattering handfuls of seed from a bag as they walked across their fields. When the young plants came up, weeds were kept away by hoeing. Ripe wheat, oats, and hay were cut with sickles and scythes.

△ *Until the arrival of modern farming methods and machines, even the simplest jobs, such as hoeing to remove weeds around the crops, were time-consuming manual tasks.*

▷ *In the Middle Ages, fields were divided into strips and each family tended their own crops.*

As long as farms remained small, there was little chance of changing these slow and inefficient methods of farming. Small farmers were content if they managed to feed their families and have a little produce left to sell at the local market. Even if equipment had been available, they did not have any extra money to spend on it. Many even had to borrow or share the oxen they used for plowing. They could not afford to risk trying out new ideas.

Huge changes in farming began when increasing demand from the growing towns and cities meant a need for more efficient farming methods. Landowners started to take back the land that they had previously leased to the small farmers, and began to cultivate it themselves. The families who had worked the small farms went to work for the landowners, although they were sometimes allowed a little land to grow food for their own tables. The landowners' large farms produced food for the growing non-farming population.

LORD OF THE TURNIPS

One step forward was to introduce changes in the way crops were grown. For many years, farmers had varied the crop on each piece of land from year to year, mainly to prevent plant diseases spreading from one year's crop to the next. From about 1700 onward, this system of "rotation" became more scientific, alternating crops such as clover and turnips, which enriched the soil as well as providing animal fodder, with wheat and barley, which were good market crops. Although farmers did not know the scientific reasons why this produced better crops, they found by experience that it did.

The English landowner who became known as the pioneer of this method was Lord Townshend (1674–1738), who owned a large estate in Norfolk. He used a four-year rotation, growing wheat, turnips, barley, and clover on his fields in successive years. Townshend was such a keen believer in the benefits of growing turnips on his land that he was given the nickname "Turnip Townshend."

Like many new farming methods, Townshend's ideas spread slowly, but by the middle of the nineteenth century turnips played their part in the cycle of crops over much of England and increasingly in mainland Europe.

CAST IRON

Meanwhile, another development had sparked off other changes in farming. In 1708, Abraham Darby (1676–1717) opened a foundry in Shropshire in England to make cast iron. Wrought iron had been used to make plows in China since about 500 B.C., and in Europe about 500 years later, but cast iron was stronger and stayed sharper.

◁ *Lord Townshend was known as "Turnip Townshend" after the method of crop rotation he advised.*

▷ *Turnips were an ideal fodder crop. Not only could they be eaten by animals and humans, but they also helped to improve the soil for growing other crops.*

The first plow using cast-iron plowshares appeared in 1730. These were so expensive that only owners of large farms could afford them.

SPEEDING THE HARVEST

These cast-iron tools were simply improvements on earlier versions. The next stage was something completely new. It was to mechanize the most expensive and time-consuming stage in farming: the harvest.

Until the nineteenth century, wheat was harvested by hand labor in which everyone in a village, from young children to old people, took part. The wheat was cut by men wielding sickles or scythes. Women and children bound the stalks into sheaves and then stood them in shocks to dry out. Women with rakes gathered up any stalks that the others had missed. The shocks were later collected and taken to a barn.

Once the wheat was ripe, the race was on to gather it in and get it under cover. Then came the long task of threshing by hand, using whiplike flails. It took about five days to thresh the wheat from half a hectare of land. Finally, the grain was "winnowed," or tossed in baskets to remove the flakes of chaff, and sorted to separate the larger seeds from the smaller.

At harvesttime, everyone helped out in the fields. Even the very youngest and oldest in the community had their jobs.

Many people worked on inventions to speed up the harvest. The first target was the tedious process of threshing.

In 1784, a Scottish inventor called Andrew Meikle (1719–1811) produced a threshing machine with a rotating drum which could be driven by wind, water, or horse power. It took less than a day to deal with as much grain as could be threshed by hand in five days.

Meikle did not make much money from his invention, and died poor, but his machine, and improved versions of it, greatly speeded up the threshing stage of the harvest. In Britain, it was so successful that farmhands, deprived of their winter work, took to breaking up threshing machines. Later, the machines became even more efficient when they were linked to steam engines.

REAPING MACHINES

Inventors next turned their attention to reaping, or cutting the wheat. A gang of five, one cutting the wheat with a scythe and the others gathering and binding it, standing it in shocks and raking up the stray stalks, could reap almost a hectare a day. A huge workforce was needed to harvest even a moderately sized farm.

A Scotsman, Patrick Bell (1799–1869), was the inventor of the first efficient

JETHRO TULL AND THE SEED DRILL

Seed drills to replace the inefficient method of sowing seed by broadcasting were used in Mesopotamia over 5,000 years ago, but in most farms in the Western world seed continued to be broadcast until well into the nineteenth century.

The first successful seed drill in the West was invented by an English lawyer turned farmer, Jethro Tull (1674–1741). He made his first horse-drawn drill, using a spring-loaded flap to control the flow of seed to the soil, in 1701. Tull was a musician, and he took the idea of the spring-loaded flap from the mechanism of a church organ. Tull's drill not only spread the seed more evenly, but the fact that it sowed in rows meant that hoeing and harvesting became easier. However, most farmers preferred to rely on the old method, and it was not until about 150 years later that drilling became the normal method of sowing.

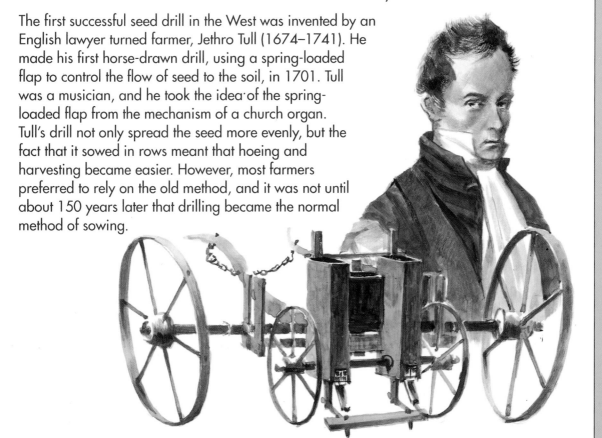

Most of the machines invented during the mid-nineteenth century to speed up the harvest were unsuccessful.

△ *In Bell's reaper, produced in 1826, the mechanism was pushed in front of the horses.*

▷ *It was not until McCormick's reaper, produced in the 1840s, that the mechanization of the harvest finally took off.*

reaper. He demonstrated it in 1826 and it won a prize, but few farmers were interested. Other inventors had the same experience. It was not until the 1840s when an American, Cyrus McCormick (1809–1884), began to sell the reaper that he had invented that the idea of mechanized harvesting took off.

▷ *Obed Hussey (1792–1860) produced his machine in 1837. In his reaper, the cutting mechanism was at the side of the machine.*

FARMING THE PRAIRIES

McCormick hit just the right moment to launch his machine. The American West was just opening up, and pioneer farmers

saw the possibilities of growing vast quantities of wheat on the prairies, if only they could harvest it quickly. But there were simply not enough people on the isolated prairie homesteads to harvest by hand. McCormick made it easy for American farmers to buy his machine by letting them pay for it over eighteen months. This way, they could buy it out of the profits of a year's harvest. The other important feature of the McCormick reaper was that it was mass-produced, using identical parts. If a part failed, it could be easily replaced.

Having sold well in North America, the McCormick reaper arrived in Europe. Soon, McCormick's factory was turning out 4,000 reapers a year.

Once reaping and threshing machines had both been invented, the next stage was to combine the two operations in one machine. That way, a machine could move into a field of standing grain and leave it with the grain already cut, threshed, bagged, and ready for market. The first horse-drawn machine of this kind appeared in the USA in the 1830s, although it was not widely used until the 1870s. It was called the

△ *Bakewell's Leicestershire sheep.*

"combination harvester," and was the ancestor of today's combine harvester.

PLOWING BY STEAM
Meanwhile, attempts were made to find a cheaper alternative to horses as a source of power on the farm. Horses had to be bred and trained to their work. They demanded feeding, care, and skilled control. After the invention of the steam engine in the 1770s, efforts were made to use steam as a source of power for farming.

Steam power could be used with Andrew Meikle's threshing machine and its successors, and threshing gangs moved from farm to farm with their machines, following the harvest. Steam power was also used for plowing, with one or two engines pulling a plow on a cable from one side of a field to the other. But only the largest and most adventurous farmers used steam to plow. True mechanization of farming had to await the invention of the gasoline-powered tractor, and later the diesel versions, in the early 1900s.

Even then, because of the cost of tractors and fuel, many small farmers went on using horses for plowing until the 1940s and 1950s.

BREEDING THE BEST
Arable farming, or the production of crops from the soil, is only one side of agriculture. The other is the raising of livestock for milk, wool, and meat. Livestock farming also made a scientific leap forward from the eighteenth century.

For centuries, farmers had bred from their best animals to produce livestock that gave more meat or wool, or that were more suited to their environment. The growing of the new fodder crops

△ *During the late nineteenth century, teams of workers would travel around the country with steam threshers, stopping to work wherever they were needed.*

such as clover and turnips led to an improvement in livestock generally, and inspired many farmers to experiment with selective breeding to improve their stock still more.

One of the great pioneers of selective breeding was a Leicestershire man named Robert Bakewell (1725–1795). He was about twenty when he began his experiments in sheep breeding.

His achievement was the Leicestershire breed of sheep. They were small-boned, barrel-shaped animals with short legs and small heads. Leicestershires fattened quickly, producing salable meat in two years instead of the usual four. Bakewell also emphasized the importance of good fodder, shelter, and handling.

SPREADING THE WORD

Leicestershire sheep were more expensive than other breeds, and Bakewell's recommendations on animal care meant that his methods could be followed only on the more prosperous farms.

These included the farms on the estates of Thomas Coke (1752–1842, pronounced Cook) in Norfolk and the fifth Duke of Bedford (1765–1802) in the English Midlands. Both of these immensely rich landowners played a part in spreading the new scientific farming ideas. Both held annual "Sheep Shearings" which were Britain's first agricultural shows, and built model farms where up-to-date methods of livestock and arable farming were demonstrated to interested visitors.

METAL FRAMES

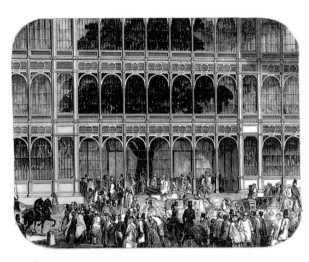

For thousands of years, buildings were made of brick, stone, and wood. Then, about 200 years ago, architects began to use metal frames to revolutionize building methods and designs, shaping the city skylines of today.

I n 1779, there was only one topic of conversation among people in the village of Coalbrookdale in Shropshire, England. Nearby, where the main road crossed the River Severn, a new bridge was being built. Its one arch crossed the river in a single span of 98 feet (30 meters).

What made the bridge unusual was that it was built entirely of iron, the first of its kind in the world. Wood or stone were the usual bridge-building materials. People wondered if the iron bridge would be safe, especially as it would carry heavy coal wagons. These worries increased when they heard that the bridge would be held together not by nuts and bolts but by a system of wedges.

THE BRIDGE HOLDS

Abraham Darby III (1750–1789), grandson of the first man to make cast iron in England, built the iron bridge and proved the critics wrong. The bridge held

△ *The Crystal Palace, scene of Great Britain's "Great Exhibition" of 1851, was built of cast-iron pillars and thousands of sheets of glass.*

△ *The iron bridge at Coalbrookdale became so famous that the town around it was renamed "Ironbridge."*

firm, and is still there today. Darby had shown that iron was an ideal material for building large structures. His bridge started a building revolution.

The bridge at Coalbrookdale was the first of many. Among those who were impressed with it was a Scottish engineer, Thomas Telford (1757–1834). In 1795, when he came to build the Pont Cysyllte Aqueduct in North Wales to carry a canal across a wide valley, he chose a trough of cast-iron plates, bolted together and mounted on stone piers, to take the flow of water. Telford chose iron again when, in 1802, he submitted a design for a new

London Bridge. This would have crossed the water in a single iron span of 600 feet (183 meters). Telford's design was not chosen, but it pointed to the future.

THE IRON PALACE

At the beginning of the nineteenth century, there was great enthusiasm for the new material. If cast iron could be used for bridges, why shouldn't it be used in buildings? In 1802, when the architect

James Wyatt (1746–1813) designed a new palace at Kew, near London, for King George III (1738–1820), he included iron in the structure. Around the same time another British architect, John Nash (1752–1835), rebuilt a house at Brighton belonging to the Prince Regent, the future King George IV (1762–1830). In the Brighton Pavilion, which can still be seen, he included iron staircases and iron columns to hold up the roofs of the larger rooms. These columns were disguised as the trunks of huge palm trees. Other designers began to use iron to support the balconies of churches.

▽ *The huge machines used in the manufacture of textiles needed to be housed in enormous rooms with high ceilings. Iron could provide the ideal skeletons for such buildings.*

THE I BEAM

In the early days of building with iron, much thought was given to the best way of shaping girders to give the greatest strength combined with the lightest possible weight.

In 1850, the Britannia Bridge across the Menai Strait in North Wales was opened. It was built of rectangular iron tubes designed by a Scottish engineer and shipbuilder, William Fairbairn (1789–1874). But rectangular tubes were expensive and heavy, so Fairbairn worked on a cheaper and lighter alternative.

His solution was the "I beam," sometimes called an I-girder or plate-girder, with its upright longer than the head and tail pieces. This gave the necessary rigidity and strength for building. I beams were first demonstrated at an exhibition in Paris in 1855.

▷ *The glass and iron structure of the Crystal Palace in London was a revolutionary design.*

▽ *Most early iron-framed buildings were given an outer shell of brick, like this one at Albert Dock in Liverpool.*

IRON FACTORIES

Before long, industry too began to make use of the new iron-framed buildings. The large, new spinning and weaving machines which were coming into use at this time were operated by belts from an overhead shaft usually driven by steam, and demanded huge, open floor areas for production and storage.

They also needed to be housed in buildings with several stories to make the maximum use of space. A system of cast-iron pillars and beams creating spacious,

open rooms was the perfect answer.

The first factory to be built in this way was a spinning mill at Shrewsbury in England, opened in the 1790s. This example was quickly followed, although the cast-iron frames were usually disguised to look like stone.

James Watt (1738–1819), whose development of the steam engine had brought about a revolution in industry, became interested in building factories from cast-iron parts made in his own factory. One factory he designed, a mill built at Salford near Manchester in 1801, was the first to use I-shaped beams. These are now standard components of large building structures.

BUILDING IN A FRAME

So far, cast iron had only been used as part of the internal structure of buildings. Stone or brick walls were still used for the outward appearance of the buildings and to help support floors and roofs. The next step was to use a metal frame for the outside of the building and let it take all the weight. For the outside walls, the frame could then be filled in with lighter material such as metal sheets or glass.

One of the pioneers in this kind of building was an American, James Bogardus (1800–1874). He became interested in architecture after inventing a wide variety of small devices ranging from engraving machines to gas meters. In 1847, Bogardus built his own five-story factory in New York which was the first cast-iron building in the United States. He went on to build many others, using standard-sized parts which could be assembled in a variety of different ways. These were the first mass-produced buildings.

GETTING TO THE TOP

As more stories were added to buildings, it became clear that people were not going to be willing to trudge up flight after flight of stairs. Mechanical hoists had been used to carry goods from floor to floor, but people were afraid to travel in them in case the ropes broke.

An American inventor, Elisha Graves Otis (1811–61), solved the problem with a safety device which held the hoist in place if the rope slipped or broke. A series of vertical pieces would catch hold of the elevator in case of an accident.

The Otis "Safe Elevator" was first demonstrated at the New York World's Fair in 1853, and was soon being installed in offices and department stores. Without the invention of the elevator, the tall buildings of today would have been impossible.

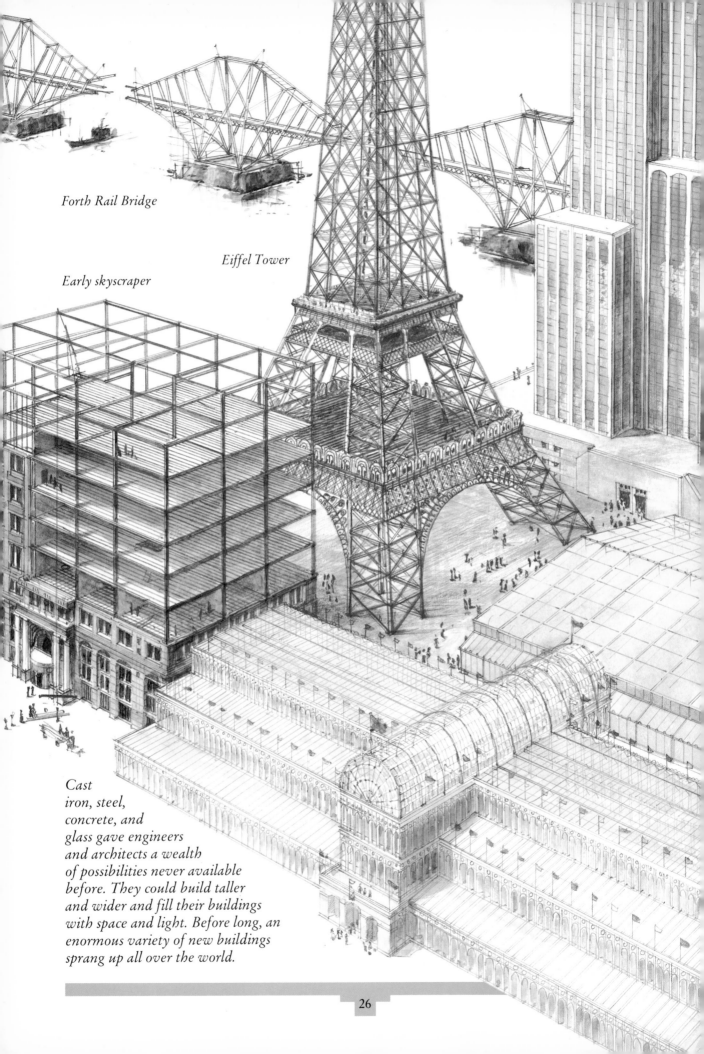

Forth Rail Bridge

Eiffel Tower

Early skyscraper

Cast
iron, steel,
concrete, and
glass gave engineers
and architects a wealth
of possibilities never available
before. They could build taller
and wider and fill their buildings
with space and light. Before long, an
enormous variety of new buildings
sprang up all over the world.

Empire State Building

*Palais des Machines
Paris, France*

*Crystal Palace
London, England*

THE RAILWAY AGE

Meanwhile, in Europe, there were other developments. The railroad had created a need for large buildings, especially at city stations. Cast-iron structures were ideal. One of the first of these large stations to be built was at Paddington in London. It was jointly designed by Isambard Kingdom Brunel (1806–59) and Matthew Digby Wyatt (1820–77) and opened in 1854. With a graceful three-arched iron and glass roof supported on slim pillars, it spanned 240 feet (73 meters).

The design of Paddington Station was influenced by an even more dramatic building which had arisen in London in 1851. This was the year of the Great Exhibition, which was to be a showplace for all the products of the British Empire. To house it, a vast glass and iron building known as the Crystal Palace was erected in London's Hyde Park.

The Crystal Palace was designed by Joseph Paxton (1801–65), who had started his working life as a gardener. This had led him to build iron and glass greenhouses, and it was on these that he based his design for the Crystal Palace. As the centerpiece for an exhibition which was acclaimed worldwide, it brought valuable publicity for the techniques of metal construction. The Crystal Palace remained in the public eye for eighty-five years. When the Great Exhibition was over, the building was moved to south London as a concert hall and exhibition center. It remained a landmark until it burned down in 1936.

THE GREAT TOWER OF PARIS

Another great European exhibition, the Paris Exhibition of 1889, provided two more striking examples of iron buildings. One, the Eiffel Tower, is still one of the most famous. Gustave Eiffel (1832–1923) had designed iron railroad bridges and station buildings before he was asked to design a centerpiece for the Exhibition. The result was a tower 984 feet (300 meters) high, then the tallest building in the world. It was built of 12,000 factory-made parts which were all numbered before being put into place. Although there were fears about the Eiffel Tower's safety in high winds, as well as a great deal of criticism of its appearance, it was to become a much-loved feature of the Paris skyline. The second great iron building at the Paris Exhibition was an exhibition hall, the Palais des Machines. This was another glass-walled and roofed building, 1,375 feet (420 meters) long.

ROLLED STEEL GIRDERS

Cast iron as a building material was by now giving way to rolled steel, which was stronger and more resistant to rust. The first rolled steel girders for building were produced in England in 1855. At first, steel girders were expensive and added hugely to the cost of building, but improved production methods lowered the price. The use of steel girders was given a boost when they were chosen for the 5,330-foot (1,624-meter) Forth Railway Bridge in Scotland, built between 1880 and 1890.

In the U.S., architects adopted steel frames with enthusiasm for the building of the first skyscrapers. The Home Insurance Building in Chicago, finished in 1885, was the world's first skyscraper, although today, with only ten stories, it looks more like a model of the real thing. It had a "skin" of thin stone walls built round a framework of cast-iron columns and steel girders. Skyscrapers became particularly popular in the U.S. because of the shortage of land in the cities. By 1910, there were more than ninety buildings in New York and Chicago with more than ten stories. By 1920, the number had grown to 450.

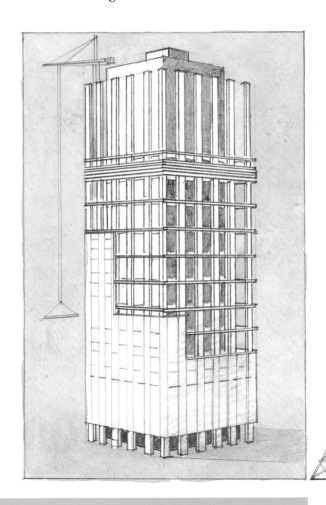

▷ *There are two main ways of building a skyscraper. Using cranes, it is built one floor at a time; using jacks, floors are assembled at ground level and then hauled into position.*

▷ *Seven of the world's tallest buildings (from far right to left):*
1. *CN Tower, Toronto, Canada (1975), 1,821 feet (550 meters)*
2. *Sears Tower, Chicago, U.S. (1974), 1,454 feet (443 meters)*
3. *World Trade Center, New York, U.S. (1973), 1,368 feet (412 meters)*
4. *Empire State Building, New York, U.S. (1931), 1,250 feet (381 meters)*
5. *Standard Oil Building, Chicago, U.S. (1973), 1,136 feet (346 meters)*
6. *John Hancock Building, Chicago, U.S. (1969), 1,127 feet (343 meters)*
7. *Eiffel Tower, Paris, France (1889), extended 1959 to 1,053 feet (321 meters)*

HIGHER AND HIGHER

Meanwhile, yet another type of construction was becoming popular. This was reinforced concrete, which was liquid concrete poured over steel rods. The first large steel-and-concrete skyscraper was the Woolworth Building in New York, completed in 1913 and designed by Cass Gilbert (1859–1934). It had fifty-two stories and rose to a height of 794 feet (242 meters). Of all the buildings in the world, only the Eiffel Tower was taller. But the Woolworth Building held the record only until 1931, when New York's Empire State Building, 1,250 feet (381 meters) high, was opened. In 1974, the record fell again to the Sears Tower in Chicago, which rises to 1,454 feet (443 meters).

The Woolworth Building, with its pencil-like tower rising through the sky, was a milestone in architecture. From then on, city skylines in America, and later throughout the world, would feature the towers of steel-framed skyscraper office blocks, hotels, and apartments, competing in their height and splendor, with all the descendants of Abraham Darby's innovative iron bridge in Shropshire.

MODERN MEDICINE

Today, doctors can prevent or cure diseases which were once major killers. Every day, surgeons carry out operations that only a few years ago would have seemed like miracles.

I n 1980, an amazing news story was broadcast all around the world. Smallpox had been conquered! There had been no new cases anywhere in the world for three years. For centuries, smallpox was a major killer disease. It spread rapidly and uncontrollably. Doctors knew of no medicine that would cure it. Not everyone who caught smallpox died, though millions did. Victims who survived were left with the scars of the smallpox sores all over their bodies,

reminding them of their suffering for the rest of their lives.

The story of how smallpox was conquered is just one example of how medical and scientific research has improved the lives of everyone living today. The good news of 1980 began about 200 years earlier in the little town of Berkeley in England. The town's doctor, Edward Jenner (1749–1823), had

△ *In the late eighteenth century, people who worked with cows often fell victim to cowpox, a disease which seemed to stop them from getting smallpox.*

heard local people say that people who had suffered from cowpox, a mild disease which was caught from cattle, seemed afterward to be protected from catching the much more serious smallpox. It was already known that people who survived smallpox, although they had to live with its scars, could not catch it again.

JENNER TAKES A CHANCE

Edward Jenner's idea was that if people could be deliberately infected with cowpox, this might protect them from smallpox in the future. In May 1796, he had the chance to try his idea out. Sarah

△ *Jenner's experiment with young James Phipps was a tremendous risk, but it paid off, saving the lives of millions of people.*

Nelmes, a local dairymaid, came to him with cowpox. He took scrapings from Sarah's cowpox sores, and used them to inoculate eight-year-old James Phipps. Now Jenner had to discover whether James would catch smallpox. In July 1797, the doctor inoculated him with smallpox.

It must have been an anxious wait to see if James developed smallpox, but Jenner's idea was proved correct. The

earlier cowpox inoculation prevented smallpox from taking hold.

Jenner's work had impressed the British royal family, and they supported him in setting up a program of smallpox vaccination in London. In the first eighteen months, from 1803 to 1804, vaccinations were given to about 12,000 people and the number of deaths from smallpox fell from over 2,000 to about 600. Soon afterward, European countries began to make vaccination compulsory, and as medical services spread throughout the world so too did the news about vaccination. The United Nations funded the last great program of vaccination in Africa that resulted in the final conquest of the disease in 1980.

HOW THE BODY WORKS

Edward Jenner's work on smallpox was part of a more scientific approach to health problems which was taking over from the superstition and ignorance of the past. Little was known about how the body works or how disease spreads. One of the first scientists to study the human body closely was a Belgian, Andreas Vesalius (1514–64). When he became a professor at Padua University in Italy, he began a study of the body by dissecting the corpses of executed prisoners. His drawings and charts identifying the functions of bones, muscles, the nervous system, and other organs were published, but he met great opposition from the Roman Catholic Church and was forced out of his job. His books remained, and became textbooks for later generations of scientists, doctors, and surgeons.

▷ *It was not until the sixteenth and seventeenth centuries, when doctors began to dissect and study corpses, that they really began to understand how the human body works.*

One of these was an Englishman, William Harvey (1578–1657), who studied at Padua where Vesalius had taught. Harvey returned to England to practice as a doctor in London, and it was there that he began to study the circulation of the blood. Before Harvey's time, no one really understood how the heart pumps blood round the body. His discoveries, published in 1628, led doctors to a completely new understanding of the body and of how diseases spread in it.

William Harvey's work was based on patient observation and experiment, and other doctors adopted his scientific methods. Harvey's message was that study led to understanding, and in the eighteenth century there was a growth in the number of medical schools in centers such as Edinburgh. Edward Jenner attended one of these, at St George's Hospital in London, from 1770 to 1773.

DETECTIVE WORK

Sometimes, advances resulted from investigations which were more like detective work than medicine. One of the greatest scourges of the nineteenth century was cholera, a disease that was almost always fatal within a few days. An epidemic of cholera spread across the world, starting in China in 1826. By 1830, it had reached Russia and then spread westward through Europe, killing millions. There were outbreaks in Britain in 1832, 1848, and 1854.

No one knew what caused cholera or how it spread. One theory was that it was carried on the wind and was therefore unavoidable. But some doctors noticed that it was most common among poor people, living in overcrowded homes with poor sanitation and communal water supplies.

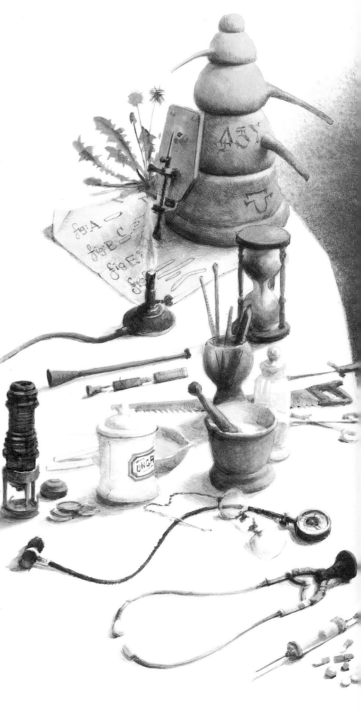

Modern medicine, with its scientific basis, seems a long way from traditional remedies, but today both forms of treatment are often used side by side.

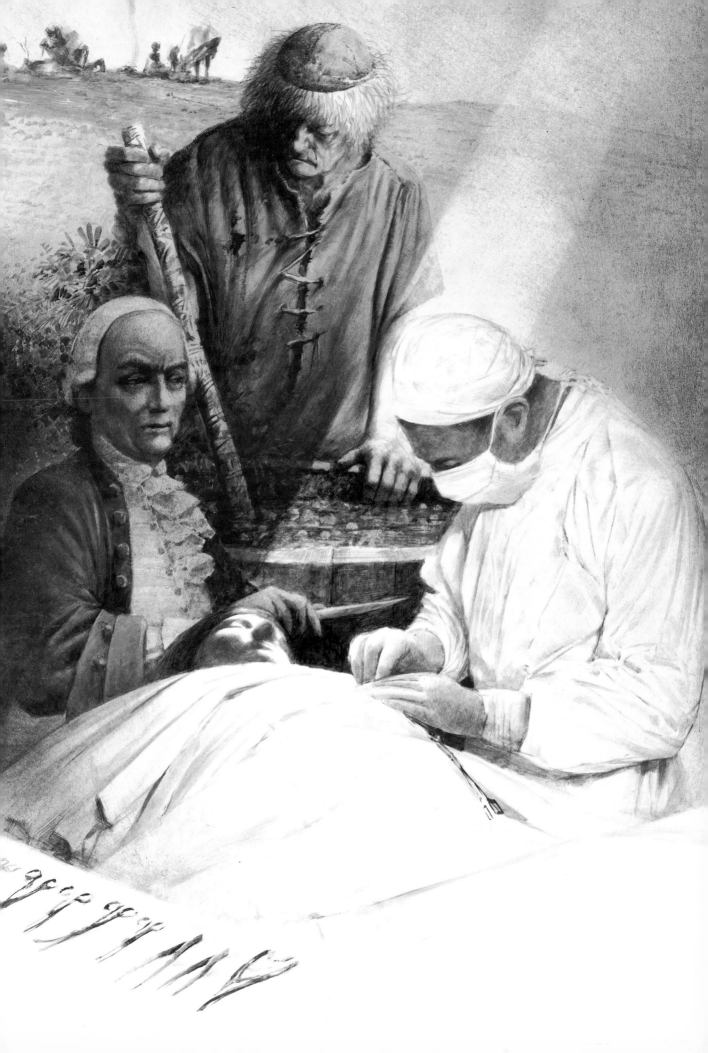

In the nineteenth century, people began to realize that the poor housing and sanitation in overcrowded cities was a major cause of disease.

The truth was proved once and for all in 1854 when an English doctor, John Snow (1813–58), painstakingly traced the source of a cholera outbreak in central London. He found that almost all the victims were members of families that took their water supply from a particular street pump. When the pump handle was removed, the outbreak began to fade. It was found later that the water in the pump had been contaminated by cholera-infected sewage. This striking example showed how the way people lived led to poor health. It was the start of improvements in public health, including better housing, clean water supplies, and the proper disposal of sewage. In many countries, these aims have not yet been achieved, and cholera is still a threat.

LOOKING AND LISTENING

Another development in the nineteenth century was the increased use of instruments to study disease and diagnose illness. The earliest of these was the microscope, invented in the Netherlands round about 1590. In the 1650s, Anton van Leeuwenhoek (1632–1723) began making microscopes which enabled him to study blood vessels, muscle fibers, and the layers of the skin. He also discovered the existence of microorganisms in human tissue. Few doctors recognized the significance of Leeuwenhoek's work at the time, but two centuries later it became the basis of a new wave of progress in medical technology.

The microscope was to prove invaluable in medical research, but there were also a number of new instruments to help the doctor's everyday work in the surgery. These helped to form a general view of the patient's health as well as to identify particular health problems.

A French doctor, René Laennec (1781–1826) was the inventor of the medical instrument we are all familiar with, the stethoscope. Laennec discovered that a rolled-up piece of paper helped him hear his patients' heartbeats more clearly, and he developed a wooden tube with a sound-collector at one end and an earpiece at the other. This type of stethoscope became popular from the 1830s until it was followed twenty years later by the instrument used today, with a flexible tube and two earpieces.

TAKING THE TEMPERATURE

The thermometer was another instrument that came into use in the nineteenth century. It had been invented over 200 years earlier, but the thermometer used until 1867 was a clumsy instrument a foot (thirty centimeters) long, which took twenty minutes to register a temperature. An English doctor, Thomas Allbutt (1836–1925), invented a shorter version which took only five minutes to register.

By the end of the nineteenth century, doctors had an instrument called a "sphygmomanometer," for measuring blood pressure. A rubber cuff was fitted round the upper arm and inflated with a rubber bulb. The pressure of air needed to stop the circulation in the arm was measured by a tube of mercury.

Instruments were also developed in the nineteenth century to enable doctors to see inside the patient's body. The first was the "ophthalmoscope," which gave a good view of the inner eye, invented by a German doctor, Hermann von Helmholtz (1821–94), in 1851. Effective instruments, called "endoscopes," to examine internal organs had to await the invention of the electric lightbulb in the 1880s. More recently, the techniques of endoscopy have been transformed again with the use of fiber optic tubes which transmit pictures to television screens from inside the body.

THE "MAGICAL" X RAY

Of all the pieces of diagnostic equipment adopted in modern medicine, one of the most important followed the discovery of X rays by a German physicist, Wilhelm von Roentgen (1845–1923). He discovered in 1895 that X rays could be used to take photographs of the inside of a patient's body. This discovery enabled doctors to study, for example, the condition of a patient's lungs or how a bone has been broken. X-ray machines were adopted with enthusiasm by doctors, and have since proved invaluable in diagnosing and

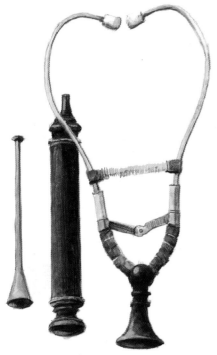

△ *Nineteenth-century stethoscopes.*

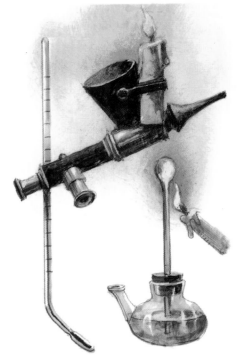

△ *Early thermometers and endoscopes.*

△ *Rapid surgery was important on the battlefields of the nineteenth century.*

treating diseases such as tuberculosis and cancer, as well as in general surgery.

ON THE OPERATING TABLE

Today, surgical operations are fairly straightforward, but it was a very different story 200 years ago. If you went to the dentist, he pulled out your infected teeth with pliers and you had to put up with the pain. More serious surgery was carried out only if there was no other way to save a patient's life. Even so, many patients died during the

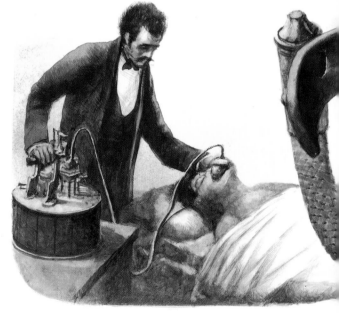

operation. There was no way of killing the pain of the operation. Patients were held down, screaming, while it was going on. All that surgeons could do to help was to try to get the job over as quickly as possible. But that was not the end. Often, the wounds from the operation became infected and the patient died of blood poisoning. The reason why infection spread in the body so easily after an operation was not properly understood.

The first improvement was the introduction of anesthetics to make the patient unconscious while surgery was being carried out. This was not only a question of making operations less painful. If the patient was unconscious, the surgeon could take more time and work more accurately. Problems such as the loss of blood during an operation could be overcome more easily. Longer operations could be carried out, and this increased the range of health problems that surgery could treat.

TREATING THE WOUNDED

Pressure to improve standards of surgery came in the nineteenth century from an unexpected source. It was a time of savage wars throughout Europe and of the American Civil War. New weapons

ANESTHETICS

The first surgeon to use ether as an anesthetic was a young doctor in Jefferson, Georgia, in 1842. Crawford Long (1815–78) removed a tumor from the neck of a patient after making him unconscious. Two years later, a Vermont dentist, Horace Wells (1815–48), used nitrous oxide, or "laughing-gas," to extract teeth painlessly.

James Simpson (1811–70) was a Scottish doctor who in 1847 started using ether as an anesthetic for mothers in childbirth. He was not satisfied, and began the search for something better. The answer he found was chloroform. Many people were opposed to the use of anesthetics in childbirth. But opposition faded after Queen Victoria (1819–1901) agreed to have chloroform during the birth of her eighth child in 1853.

Nineteenth-century ether inhaler.

LOUIS PASTEUR

Most nineteenth-century surgeons thought that their patient's deaths were caused by a mysterious poison in the air. Understanding of disease took a huge leap forward when Frenchman Louis Pasteur (1822–95) worked out his "germ theory" showing that infections were caused by microorganisms called bacteria.

Pasteur was trained as a chemist, and in 1854 he was appointed dean of the faculty of sciences at Lille University. He began to study microorganisms as part of an investigation into why wine went sour and food went bad. He discovered that bacteria which cause decay in food can be killed by heating it. This process, still used to stop milk going sour, is called "pasteurization" after its inventor. It was only afterward that he went on to apply his findings to human medicine.

Later, Pasteur developed a vaccine against anthrax, a deadly disease passed from cattle to humans, and, perhaps most spectacularly, a vaccine for rabies, which is passed on if someone is bitten by an infected animal, such as a dog or a fox.

meant that more and more soldiers were wounded on the battlefield. Many were left to die, and even those who were given medical help did not survive long. It was this situation that led to the foundation of the Red Cross movement after a particularly cruel battle between French and Austrian troops at Solferino in northern Italy in 1859. It led also to the realization that many of the lives of the wounded could be saved if their injuries were treated immediately, before fatal infection set in. But how could infection be stopped?

KEEPING CLEAN

The pioneer work of the French chemist Louis Pasteur on his "germ theory" showed that infection was caused by bacteria, which were micro-organisms similar to those observed by van Leeuwenhoek many years before. A British surgeon, Joseph Lister (1827–1912), began to apply Pasteur's theory to the operating theater at Edinburgh Medical School, where he was a professor of surgery.

Lister insisted that the surgery and its equipment must be scrupulously clean. He also made use of the discovery by another professor, F. Grace Calvert (1819–73), that phenol, or carbolic acid, was effective in slowing the process of decay. Lister used a spray to create a mist of carbolic acid around patients' wounds during operations, and also used carbolic acid to clean and dress the wounds. The result was a dramatic cut in the number of patients who suffered from infection after operations. This also meant that many limbs which would previously have had to be amputated could be saved.

PREVENTING INFECTION

The lasting importance of Lister's methods lay in his insistence on germ-free conditions in the operating theater. The use of carbolic acid was soon abandoned because it was found to irritate patients' wounds.

Alternative ways were found of preventing wounds from becoming infected. Surgeons, who had previously worn ordinary clothes to carry out operations, began to wear gowns, masks, and caps. These could be sterilized after use under steam pressure in an "autoclave," which first came into use in hospitals in 1886. It was similar to a pressure cooker and based on a much earlier invention by a French scientist, Denis Papin (1647–1712). The autoclave was also used to sterilize dressings and the surgeon's instruments.

△ *Reformer Florence Nightingale (1820–1910) helped to improve the appalling conditions in nineteenth-century hospitals and founded the modern nursing profession. Her patients gave her the nickname the Lady with the Lamp because she would wander the wards at night with her lamp, checking to see that all the patients were comfortable and had everything they needed.*

Another improvement in operating techniques came in 1890, when an American surgeon began to use thin rubber gloves which could be thrown away after the operation.

Joseph Lister and his carbolic acid spray.

SPARE PARTS

There were improvements during the nineteenth century even for those patients who had to have limbs amputated. "Wooden legs" had been used for centuries, since at least 5000 B.C., but until the nineteenth century they were crude, unsightly, and had no joints. There was an important change in 1815 when an artificial leg jointed at the knee and ankle was produced for the Marquis of Anglesey (1768-1854), who had lost a leg at the battle of Waterloo.

By the end of the century the improved limb was available not only to the rich. There was a similar improvement in artificial arms, harnessed to the body and jointed at the elbow.

FALSE TEETH

Alongside improvements in medical care for the seriously ill, the nineteenth century also saw more attention for more minor health problems such as tooth and eye disorders. The invention of anesthetics made a visit to the dentist a far less painful experience, and new technology brought benefits to people who had to wear false teeth.

False teeth, made of ivory and held together with gold wires and plates, had been available for centuries for those who could afford them. But they were fitted by taking measurements inside the mouth, which was difficult to do at all

◁ *Before anesthetics, dentists wrenched out rotten teeth using pliers. Often the patients got very drunk to try to dull the pain.*

▽ *Early false teeth were not just uncomfortable, they were expensive too. The teeth were made from ivory and the wires from gold.*

accurately. In the eighteenth century, dentists began to take wax casts of the teeth. Plaster molds were then made from the casts and used as a pattern, and this improved the fitting. But the real breakthrough, which made false teeth a possibility for poorer people, came in 1844 with the invention by Charles Goodyear (1800–1860) of vulcanite, a kind of hard rubber. This proved to be a cheap and comfortable material for the dental plate, to which porcelain teeth were attached.

TOOTHBRUSHING

The causes of tooth decay were not properly understood until late in the nineteenth century. People expected to lose their teeth quite early in life, often in their thirties, and it was normal for them to have complete sets of false teeth, top and bottom. Many regarded their natural teeth as a nuisance and a source of pain which was best avoided by replacing them with false ones. In 1884, research in Berlin, Germany, showed that tooth decay is caused by bacteria feeding on acids in the mouth. This discovery led dentists to emphasize the importance of regular brushing after meals, especially for children whose teeth are still developing. Although it has taken a century to achieve, the teeth of people in the Western world are more healthy and less prone to decay than they have ever been.

TREATING POOR EYESIGHT

Poor sight was another health problem that most people believed just had to be accepted. Spectacles had been used for centuries, but they were expensive and were little more than magnifying lenses mounted in frames. In the nineteenth century, opticians began to test the sight of each eye with frames containing trial

△ *Spectacles were usually just magnifying glasses of different strengths, not made accurately for each particular patient's vision, as they are today.*

lenses. In this way, pairs of spectacles could be prescribed in which the particular sight problem of each eye was corrected. The invention of the ophthalmoscope also enabled the interior of the eyes to be examined for more serious defects.

ACHIEVEMENT

By the end of the nineteenth century, doctors had a far better idea of what causes disease, and how it can be prevented or cured, than they had a hundred years earlier. Fewer diseases were fatal, surgery was no longer so risky, and patients suffered less pain. The stage was set for even more hopeful developments in the twentieth century, but we have all benefited from the work of the scientists, researchers, and doctors of the nineteenth century.

MACHINE POWER

The story of machines began thousands of years ago when ancient people devised easier ways to lift heavy weights, split building stone, and move objects from one place to another.

F rom the simplest man-powered machines, such as the wheelbarrow or the pulley, to complex powered machines, such as the car, all machines are tools that make work easier.

We know from prehistoric sites such as Stonehenge in England, built around

△ *Wooden levers can be used to help raise a heavy weight.*

◁ *The simplest machines, such as wedges for splitting rocks, mean that strenuous tasks can be performed with less effort.*

3000 B.C., that amazing feats of moving and lifting were carried out using only human and perhaps animal power. No one is sure exactly what methods were used to carry the huge blocks of stone necessary, but rollers, levers, and wedges must have played a part. Even with the help of these devices, building projects such as the great cities of ancient Persia and Egypt took thousands of people many years to build.

WEDGES FOR BUILDING
The ancient Egyptians used wedges in a wide variety of ways. In building, they

Machines did not only help in manufacturing and building work. This catapult is a type of lever. It was used to throw burning objects over a wall during a siege.

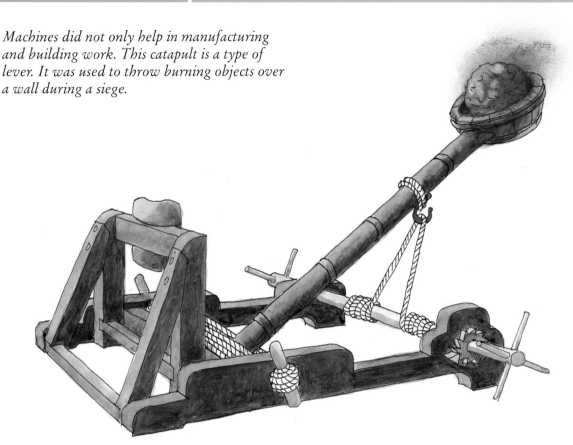

often trimmed a piece of stone to a wedge shape and then drove it between other stones to make a tight fit. They used sloping wedges as ramps for pushing stone into place. In raising stone columns, or "obelisks," they lowered the base into position down a slope, and then used ropes tied around the top to haul the obelisk upright.

The Egyptians also used wedges to split stone. They drilled holes in blocks of stone and then hammered in wooden wedges. Water was poured on to the wood, which made it expand, splitting the stone.

THE POWER OF THE LEVER

The lever was another of the simple machines available to ancient people. We use levers many times every day without realizing it. Every time we open a door, cut with scissors, or play on a seesaw, we are making use of levers. Although ancient peoples did not understand the

principle of how levers work, they used them for many tasks. The oar is a kind of lever, as is the wheelbarrow.

In the third century B.C., the Greek scientist Archimedes (287–212 B.C.) began to experiment with levers and to figure out how to make them do the most work for the least effort. Archimedes wrote that if he were given a lever long enough and a place to stand, he could move the whole Earth. This was just a fantasy, of course, but it suggested the lever's power.

It was soon discovered that the feet could operate levers in the form of treadles, leaving the hands free for other tasks. Treadles were used to operate hammers for pounding rice into flour, and in weaving machines. Most of the devices we use in the modern world, from cars to electric light switches, contain levers. The clutch, brake, and accelerator pedals of a car are all treadle-type levers.

ROUND AND ROUND, UP AND DOWN

Changing a rotary, or round-and-round, effort into a lifting or pulling effort was a challenge that the machine makers of the ancient world faced, and beat.

Although we know the device described by Archimedes for raising water as the "Archimedes screw," he was writing about a machine that he had seen rather than one he had invented. The screw can be thought of as a slope arranged in a spiral. When the handle of the Archimedes screw was turned by human or animal power, the thread of the screw carried water up the shaft.

In a similar way, turning a screw to fasten two pieces of wood converts the rotary motion of the screwdriver into a linear motion which pulls the two pieces of wood together. When gear wheels are cut so that they mesh together, they have the same effect of changing the direction of effort.

SCREWING IT UP

The screw is another ancient device. We think of it today mainly as a way of fastening wood or metal, but the screw as a fastener did not become widespread until the eighteenth century. Until then, it was not possible to make screw threads accurately enough.

The ancient world used the screw in quite different ways. A screw is, in fact, a slope, or inclined plane, which winds around in a spiral. Archimedes described such a screw device for raising water to irrigate fields, but this had been in use long before his time. The Greeks and Romans used screws in presses to extract juice and oil from grapes and olives.

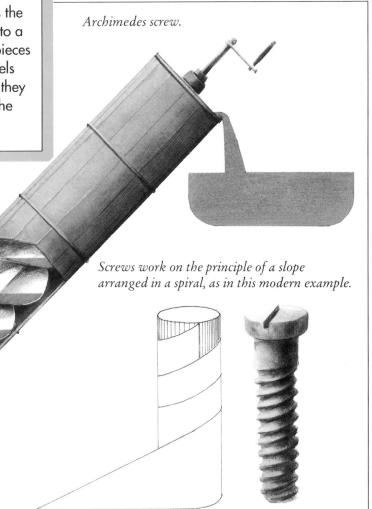

Archimedes screw.

Screws work on the principle of a slope arranged in a spiral, as in this modern example.

THE PULLEY

In a pulley, a rope lifting a load is passed between two or more grooved wheels. Passing the rope over one pulley makes the task of lifting easier, but using two or more is even more effective. The ancient Greeks are known to have used a system of pulleys to haul rock from a silver mine.

The crane was a development of the pulley. It had a tall tower called a "jib" which carried the pulleys. The jib was mounted on a platform which could be revolved, and worked by a treadmill operated by human power. The crane was possibly first developed by the Romans as a machine for loading and unloading cargoes from ships.

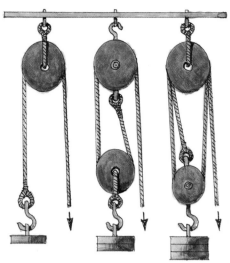

The more pulleys there are, the less effort it takes to lift a load.

THE WINCH

Another hauling and lifting machine invented in early times was the winch. A rope tied to a load passed over a horizontal wooden roller which was operated by a handle. As the handle turned, the rope wound around it, dragging the load nearer. A winch was used in a copper mine in Austria around 600 B.C. A similar device with a vertical roller is called a windlass. The ancient Egyptians used a windlass operated by ox power to lift water from rivers in jars and transfer it to irrigation channels.

After the collapse of the Roman Empire, there was no large-scale building in Europe for about 500 years. There were no longer vast armies of slaves to do the lifting and pulling. However, in the late Middle Ages, the Christian church began to build new great cathedrals throughout western Europe. All the skills of engineers and builders were called upon, and the result was a revival of machines used in the past to handle the huge blocks of stone and heavy timbers that were needed.

SEA TRADING

When it was seen how effective these machines were, they found other uses, particularly at seaports where the windlass became a valuable tool for pulling ships in close to the quays, and the crane for loading and unloading. At a time when trade by sea was increasing rapidly, machines that helped shipowners to turn their ships around quickly were welcome. Ports with such equipment as cranes and winches attracted the trade, and their merchants grew prosperous alongside the shipowners who used them.

THE FIRST MACHINE TOOL

Meanwhile, simple machines had been developing in a completely different direction, in the making of machine tools. A machine tool is a device which helps to speed up repetitive but skilled tasks, such as cutting or shaping metal or wood.

The first machine tool was the potter's wheel. Before it was invented about 6,000 years ago, pots were made either by pressing clay over a rounded object or by building up a coil of clay and smoothing together the edges. The wheel, turned by a handle or a treadle, enabled pots to be "thrown" by the fingers, building them up from a ball of clay as the wheel turned.

THE LATHE

One of the most basic machine tools, the lathe, works in much the same way, by working on an object as it turns. We do not know when the first lathes were used. A wooden bowl made using a lathe has

been found dating from about 700 B.C. Earlier Egyptian paintings show a kind of lathe being used. A cord wrapped round the object being turned is being pulled by one worker while the other uses a chisel to shape the work.

In the Middle Ages, pole lathes were often used. The cord was attached to the top of a springy wooden pole and wound round the work to be turned. At the other end of the cord was a treadle. Foot pressure from the treadle and tension from the pole kept the work turning. The advantage of this design was that only one person was needed both to operate it and to do the shaping.

▽ *The winch combines the use of a roller with a lever to make lifting or pulling a heavy weight easier. Here a pair of Greeks are using a winch, coupled with a lever and a treadle (or foot lever) to dredge mud out of a harbour. It must have been very hard work!*

THE QUEST FOR PERPETUAL MOTION

By the late Middle Ages, the development of machine technology had gone about as far as it could go while the effort that went into operating machines had to come either from humans or from animals such as oxen or donkeys. Wind and water power could be harnessed for some purposes such as grinding grain, but many operations had to be carried on continuously, whatever the weather or the flow of a river. There were now no huge armies of slaves to provide human power, as there had been in ancient Egypt or Rome, and even beasts of burden had to be fed and looked after.

This problem teased at the minds of inventors and scientists. Their understanding of the science of mechanics was not very clear. Many people tried to invent machines that, once started, would just keep going without any further energy being applied, called "perpetual motion" machines. Such a device, they thought, could be used to operate levers, winches, pulleys, and other simple machines for as long as it was needed.

They were wasting their time. Forces over which we have no control, such as friction and gravity, mean that any device with no external power source will eventually slow down and stop. Perpetual motion was not the answer to the problem of power, but it was not until the late eighteenth century, with the development of the steam engine, that a solution was found.

Here are some of the designs for perpetual motion machines.

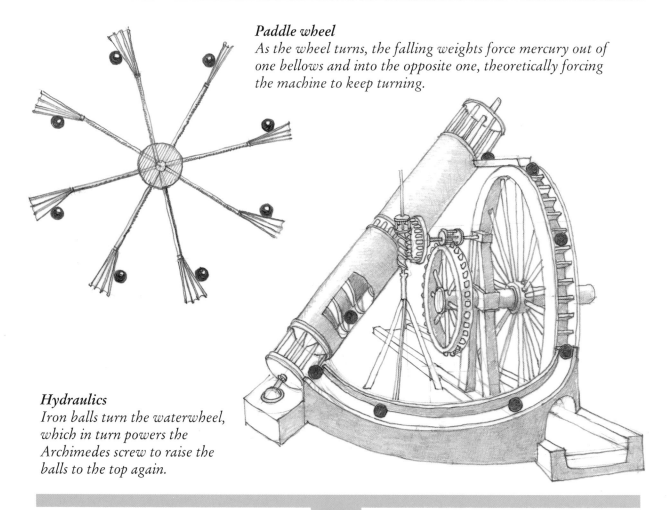

Paddle wheel
As the wheel turns, the falling weights force mercury out of one bellows and into the opposite one, theoretically forcing the machine to keep turning.

Hydraulics
Iron balls turn the waterwheel, which in turn powers the Archimedes screw to raise the balls to the top again.

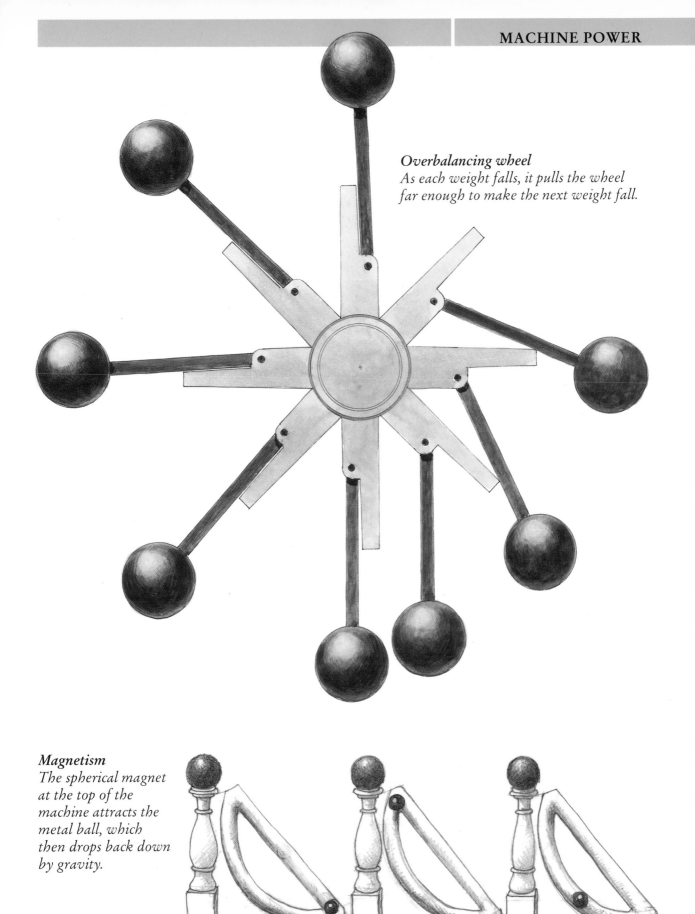

Overbalancing wheel
As each weight falls, it pulls the wheel far enough to make the next weight fall.

Magnetism
The spherical magnet at the top of the machine attracts the metal ball, which then drops back down by gravity.

The power hammer was a machine tool that reached Europe round about A.D. 1100, although it had been used in China at least 1,000 years before. It consisted of a heavy weight which was raised to a height above an anvil on which the object to be worked was placed. The weight was then released and allowed to drop by force of gravity. The power hammer saved a great deal of effort when large numbers of similar items, such as sickle and knife blades, were needed. The first power hammers were driven by waterwheels.

DAWN OF A NEW AGE

At last, in the eighteenth century, engineers came up with something the world had been waiting for: a new way of powering machinery. The need for a breakthrough had become urgent, especially in Britain where there was a fuel crisis. Wood, either dried and burned or in the form of charcoal, had been Britain's main source of fuel, but the forests were being burned faster than they could be replanted.

As a result, people turned to coal, but this meant sinking pits deeper to levels where underground water flowed. The "gin," a kind of windlass powered by horses, was used to drain the pits, but it did not work fast enough. Against this background, the steam engine was perfected by the Scottish engineer James Watt (1736–1819). Here was a source of power that could be used as a pumping engine, to help produce the fuel it

▷ *During the Middle Ages, machines began to make a difference to people's working lives. The machines used on this building site look basic to modern eyes, but they made possible building tasks which otherwise could only have been done with a huge army of workers.*

△ *Early steam engines were inefficient, but before long, steam revolutionized most industries.*

burned. Steam engines could also drive shafts that in turn could power all kinds of machine tools.

One of the problems with early steam engines was the fit between the piston and the cylinder in which it moved. If the piston was too loose, vital pressure was lost and the engine performed inefficiently. Attempts were made to seal the gap with rope or leather, but these seals quickly wore out.

CANNONS AND CYLINDERS

The solution to this problem was worked out by an English iron founder, John Wilkinson (1728–1808). His main business was making armaments, including cannon barrels. Here too, there was a problem of fitting. Barrels too large for the cannonballs did not fire accurately. Barrels that were too small were likely to blow up when fired.

In 1774, Wilkinson invented a machine which gave his cannon barrels perfectly circular bores along all their length. A

hard cutting tool on the end of an iron bar was driven round inside the barrel at high speed. The next year, Wilkinson offered to make steam engine cylinders, using the same machine, for Watt, who had just set up a business making steam engines in partnership with Matthew Boulton (1728–1809). Wilkinson's machine was just as successful in making steam engine cylinders as it had been with cannon barrels. Wilkinson cylinders were used in the hundreds of Boulton and Watt steam engines which provided the power in the new factories springing up at the start of the Industrial Revolution.

ENERGY ON TAP

Once steam engines were available, manufacturers were quick to explore ways of using their tremendous and continuous power, a source of energy that had never been known before.

Henry Maudslay (1771–1831), an Englishman, invented an accurate screw-cutting lathe. In the past, large screws had been forged and then filed by hand to make the thread, while small screws were cut entirely by hand. This was slow and highly skilled work, and as a result screws were expensive. Maudslay's lathe meant that screws could be cut easily and accurately, with threads of standard sizes.

Maudslay's factory in London produced many other machine tools, but probably his most important other invention was the "micrometer" for measuring machine parts with precision. This made use of the ability of his lathe to produce regular and accurate screw threads. The basis of the micrometer was a long screw with twenty threads to each centimeter, and pointers which could be moved by a thumbscrew up and down

The new mechanical lathe worked in much the same way as the pole lathe of the past, but was easier to control and more accurate.

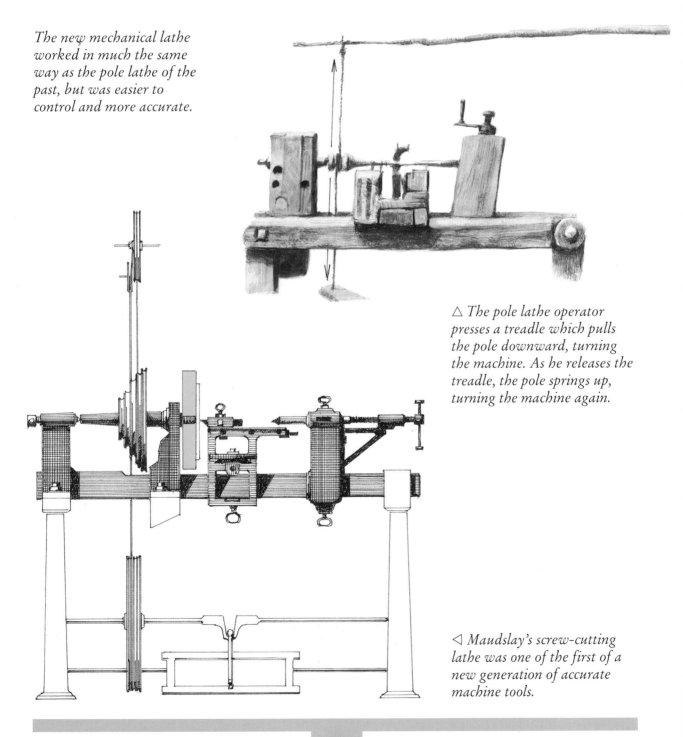

△ *The pole lathe operator presses a treadle which pulls the pole downward, turning the machine. As he releases the treadle, the pole springs up, turning the machine again.*

◁ *Maudslay's screw-cutting lathe was one of the first of a new generation of accurate machine tools.*

the threads. Using it, engineers could achieve a degree of accuracy unknown before.

STEAM PRESSURE

Scottish inventor James Nasmyth (1808–1890) invented the steam hammer in 1839, in which a weight was raised by steam pressure and then dropped on the piece of iron to be forged. The hammer was designed to forge a shaft for a new ship. The ship designer changed his plans, and the shaft was not needed, so this first Nasmyth hammer was never built. But the drawings were seen and copied by a French iron founder, who built it at his works at Le Creuzot. Nasmyth went on to develop an even better steam hammer, which used steam to control the fall of the weight as well as to raise it. The controls of the downward fall were so delicate that the hammer could be brought down on an egg so gently that the egg was cracked, but not broken. This trick was often performed at Nasmyth's works to entertain and impress visitors.

The nineteenth century saw a flood of new machine tools for mechanizing a wide range of tasks. Rolling mills were designed for producing sheet metal. Punching machines and presses were built for such work as making coins and

medals. Other machines bent and formed metal into shapes. James Nasmyth invented machines for planing and cutting slots in metal. Mechanical saws and grinders, gear cutters, wheel cutters, and many others added to the list of machine tools that produced work with a speed and accuracy never seen before.

THE WORLD'S WORKSHOP
Sparked off by the steam engine, the Industrial Revolution first took hold in Britain, and for many years many of the major steps forward were made by British engineers and inventors. Britain became known as "the workshop of the world," and talented engineers from other countries were attracted by the opportunities for progress. For example, Marc Isambard Brunel (1769–1849) was a Frenchman who began a new career in England at the age of thirty. He worked with Henry Maudslay on the development of machine tools for naval dockyards and later became a pioneer in underwater tunneling.

Brunel's son, Isambard Kingdom, (1806–1859) was one of the great railway builders. French, Belgian, and German manufacturers were keen to adopt British methods and machines, so much so that for a time the British government banned the export of machinery to Europe to protect manufacturers at home.

Britain's leadership of the industrialized world was not to last. First the United States and later Germany began to catch up in the branches of industry where Britain had excelled. Both countries were also quicker to explore the possibilities of two new sources of power

◁ *Nasmyth's steam hammer could be so accurately controlled that it could crack an egg without breaking it.*

OUT OF WORK
The development of machine tools was good news for manufacturers, who could make their products in larger numbers, and for consumers, who could buy such things as clocks, furniture, cutlery, and gardening tools more cheaply. It was not such good news for the skilled craft-workers who had previously made these things by hand.

Many such people, in trades like spinning, weaving, metal grinding, and cabinetmaking, were thrown out of work by steam-driven spinning machines, looms, and lathes. Their skills were not needed any more, and the only work they could find was as machine operators in the factories, at lower rates of pay. One of the by-products of the Industrial Revolution was to make industrial workers less satisfied with their jobs and more resentful of the people who employed them.

which were eventually to take over from steam: electricity and the internal combustion engine.

American inventors such as Thomas Alva Edison (1847–1931) and Germans such as Ernst Werner von Siemens (1816–1892) were the leaders of the electrical revolution. The internal combustion engine was the product of German engineers Gottlieb Daimler (1834-1900) and Carl Benz (1844–1929), with Rudolf Diesel (1853–1913) contributing the engine that still carries his name. But these inventors of the machines of the modern age depended on the quest for quality and accuracy in machine tools by the engineers who preceded them.

NATURAL POWER

Blowing wind and flowing water are sources of energy that need no fuel and will never be used up. When people learned to tap this natural energy, they could begin to work more efficiently than ever before.

For thousands of years, people had only the power of their own muscles to help them with physical work such as carrying water, plowing, and building. As the first civilizations developed around farming communities and the first cities were built, there was an urgent need for human labor. Many early civilizations met this need by setting up systems of slavery.

One major purpose of early wars was to bring back slaves from the conquered territories. Slave labor was responsible for the great building projects in ancient Egypt, Greece, and Rome.

STEP BY STEP

Some machines were devised to make the best use of human effort. One of these was the treadmill, an open wheel fitted with steps and connected at the center to a shaft. As slaves walked continuously inside the wheel, the shaft rotated and

△ *For thousands of years the only sources of power were the strength of a man or an animal.*

▷ *Slaves were valuable in the ancient world, because they were an important source of power.*

58

THE NORIA

The "noria" was one of the earliest types of waterwheel, which was probably used in the lands around the Mediterranean before the Roman Empire began to expand. The noria was a device for raising water from a river so that it could be channeled along ditches to irrigate the fields. It was a undershot waterwheel with paddles to catch the flow of water and jars fitted round the circumference. As the wheel turned anticlockwise in the flow of the water, the jars were filled. At the top of the cycle, the jars were emptied into the drainage channel.

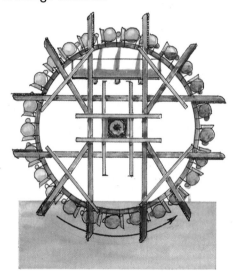

could be used to drive machinery such as millstones and lifting devices.

Ox power was also available, but its use was limited to simple tasks such as pulling plows and operating water pumps. People had not yet started to use horses, so for most work people had to rely on themselves or their slaves.

There was a breakthrough during the first century B.C., when water was used for the first time as a source of energy for grinding grain. The first waterwheels had scoops or blades fitted to the end of a vertical shaft. The top of the shaft passed through the center of the lower of a pair of millstones and was attached to the upper stone. The blades at the other end were turned as river water flowed past them. Each complete turn of the shaft produced one complete turn of the upper millstone. This simple form of water mill was first described by Greek writers, and is often called the Greek mill.

THE ROMAN WAY

The Greek mill worked, but it was not very efficient. It depended on a regular flow of water, so that any seasonal variation caused problems. But improvements were on the way. The Romans had become experts in the use of water. They revolutionized the design of water mills by turning the wheel on its side and connecting it to the millstones using gears. In this way, the stones could be made to turn up to five times for each turn of the waterwheel. The Roman design, with blades set across the wheel, made the maximum use of flowing water.

The Roman waterwheel was "undershot," which means that the water flowed beneath the blades. Undershot wheels had the disadvantage that if the river level dropped, the wheel turned only slowly, or not at all. Later, people realized that if the water actually fell as it passed through the wheel, the power would be increased. The result was the "breast" wheel, in which the water met the blades or buckets halfway down. The "overshot" wheel, where the water flowed in at the top of the wheel, was a still later development. By this time, people had discovered how to provide a more regular supply of water by building a dam upstream and channeling the water to the wheel along a ditch or pipe.

It was only in the later days of their

empire that the Romans built water mills in large numbers, but some of these were very elaborate arrangements. At Barbegal, near Arles in France, they built a series of sixteen wheels, each of which drove two pairs of millstones, making use in turn of the same flow of water. This mill could produce twenty-eight tons of flour a day, enough to feed 80,000 people. The local population needed only a fraction of this, and the rest was probably sent away to feed the Roman army.

WATER FOR EVERYTHING

By the Middle Ages, water mills were being widely used in Europe, not only for grinding grain but also for powering saws, pounding rags to make paper, driving hammers for metalworking, hoisting stone and coal from quarries and mines, tanning leather, and treating cloth.

Ownership of a grain mill was a profitable business. A landowner with a suitable river would build a water mill that farmers could use to grind their grain. The fee they paid

was usually a proportion of the flour produced, so that the owner had flour to sell without the trouble and expense of growing the grain himself. Monasteries were often the owners of large estates and they too went into the milling business.

There was a huge growth of water mills in the late Middle Ages. In the Aube district of France, there were only

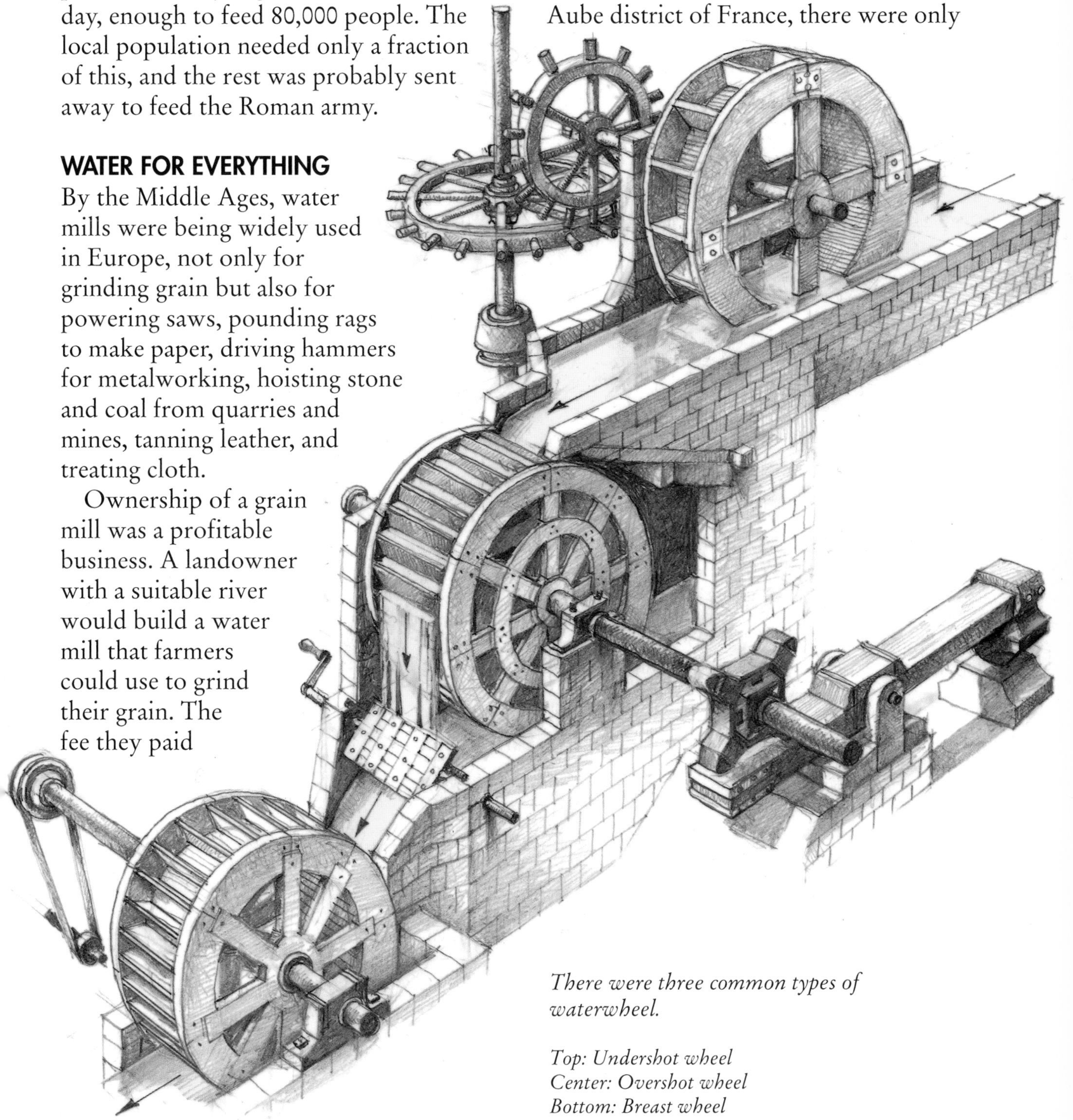

There were three common types of waterwheel.

Top: Undershot wheel
Center: Overshot wheel
Bottom: Breast wheel

fourteen mills in the eleventh century. Less than two centuries later, there were 200. For communities fortunate enough to have been built on a fast-flowing river, water mills were the key to prosperity and growth. For example, by the sixteenth century, the Italian city of Bologna at the foot of the Apennine Mountains had mills for grinding grain, metalworking, spinning, sawing, sharpening, and polishing.

CATCHING THE WIND

Across the plains of Europe and central Asia and in the Middle East, fast-flowing rivers are few and far between. These were areas where the windmill became popular.

It was not until about A.D. 1100 that windmills were first built in Europe, but they had been used much earlier in China, Afghanistan, and the Middle East.

▷ *Animals, the wind, and water were sources of power that dominated the landscape for thousands of years.*

▽ *Gearing was needed to turn the motion of the waterwheel from vertical to horizontal.*

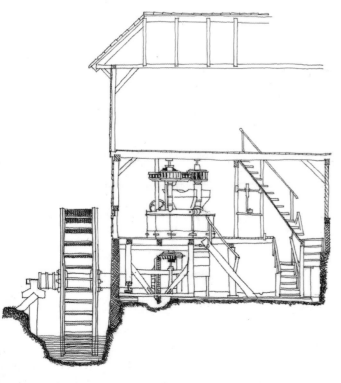

The fact that the sails of early windmills often copied the local pattern of ships' sails, triangular in the Mediterranean and rectangular in China, suggests that the idea of using wind power on land may have come from its use at sea. In the first windmills, the sails traveled around in a horizontal path to drive a vertical shaft. An early example of this type of mill can still be seen at Seistan in Afghanistan, and similar designs are still in use in remote parts of the Middle East today. The Chinese, who used their windmills for irrigation, built the frames of bamboo so that they could be moved from place to place wherever irrigation was needed.

A traditional windmill from the Middle East.

WINDMILLS IN EUROPE

The idea of using wind power may have traveled from China or central Asia with traders returning from the Far East, or possibly through Russia and Scandinavia. However, it seems more likely that windmills in the West developed independently, as their design was, from the start, quite different from those in the East. The European windmill had a horizontal shaft driven by sails which moved in a vertical path. This first type of windmill was called a "post mill." The whole structure, built of wood, could be turned by human or animal power so that the sails faced into the wind.

The spread of windmills throughout Europe took place very quickly. Windmills could be built almost anywhere, although open hilltop sites were best. Building them required no great skill. Every community could have its own mill, and not have to wait to have its grain ground at a water mill which might be miles away. Soon after the start of the twelfth century, there were hundreds and then thousands of windmills at work all over Europe.

The windmill was an invention that would last. For about 700 years, most people in Europe ate bread made from flour that had been ground by the wind. Going back a hundred years, few pictures of a rural landscape in Europe did not include at least one windmill. Until about sixty years ago, there were many windmills in Europe still grinding corn. Even today, there are still a few at work, although they are now usually tourist attractions.

THE MOVING CAP

Improvements in the design of the windmill were introduced as time went on, though some communities continued to build the simple post mill. But for most millers, the effort of constantly pushing the post mill into the wind was too much hard work. The "tower" or "cap mill" made life easier.

The tower mill had a more substantial main structure which could be made of brick or stone. This contained the millstones and heavy machinery. The sails were attached to a wooden cap which was lighter and so easier to move into the wind. The effect of this was to raise the sails higher from the ground, catching the

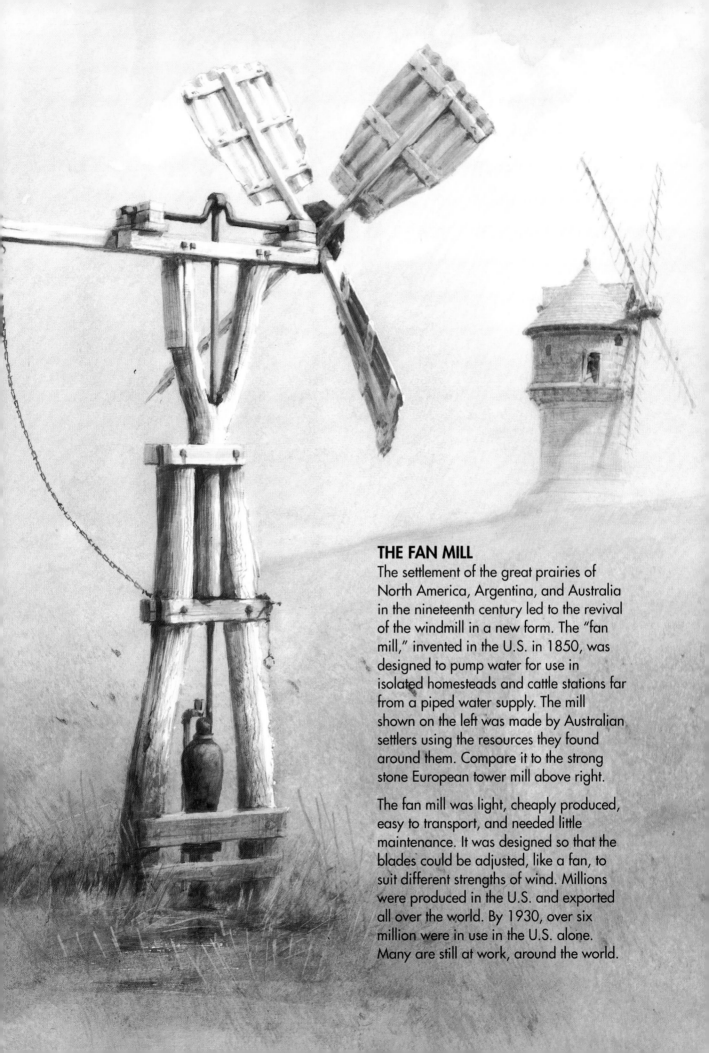

THE FAN MILL

The settlement of the great prairies of North America, Argentina, and Australia in the nineteenth century led to the revival of the windmill in a new form. The "fan mill," invented in the U.S. in 1850, was designed to pump water for use in isolated homesteads and cattle stations far from a piped water supply. The mill shown on the left was made by Australian settlers using the resources they found around them. Compare it to the strong stone European tower mill above right.

The fan mill was light, cheaply produced, easy to transport, and needed little maintenance. It was designed so that the blades could be adjusted, like a fan, to suit different strengths of wind. Millions were produced in the U.S. and exported all over the world. By 1930, over six million were in use in the U.S. alone. Many are still at work, around the world.

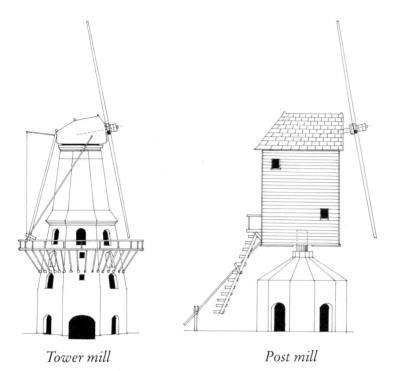

Tower mill   *Post mill*

Britain and other countries. The Dutch were also the first to use the wind to power sawmills, and the ease with which sawed timber could be produced resulted in the typical Dutch style of building in wood. The windmill had become the basic machine for many processes, from grinding exotic spices for the tables of the rich to grinding chalk to make whitewash for the walls of cottages. Even after the invention of the steam engine, it was hard to beat the economy and simplicity of wind power for these processes.

wind better and also allowing other buildings such as storerooms and the miller's house to be built alongside. If the sails went out of control in high winds, as they often did, only the cap and not the whole mill would be damaged.

An important improvement was introduced in England in 1745 when a fantail was added to a windmill. This was a smaller sail fitted opposite the main sails. The fantail caught the wind and automatically turned the main sails to face in the right direction. In countries like Britain, where the wind can change direction several times in the course of a day, this was a great help to the miller.

PUMPING AND SAWING
As with the water mill, it was soon found that windmills could perform a wide range of tasks. In the fifteenth century, the Dutch began to use windmills to pump water away from low-lying land. Later, they used them to reclaim new land from the sea, a method that was copied in

THE RETURN OF THE WINDMILL
In many parts of the world, wind and water are still the most easily available sources of energy. Electricity has been generated from the flow of water through turbines in hydroelectric dams for almost a hundred years, but until recently the possibility of using wind power was ignored. Wind generators, modern versions of the windmill, designed to generate electricity from rotating blades, have been brought into use and are supplying energy to national supply systems as well as isolated communities. It may be that we will once again come to value the source of energy that performed so many tasks for hundreds of years.

▷ *Today, natural power is becoming a valuable resource as people realize the polluting dangers of fossil fuels. Wind, water, and solar power have all been developed to produce electricity.*

ELECTRIC POWER

For thousands of years, people wondered about the mysterious lightning they saw in the sky. Their questions revealed the existence of an energy source which is now available at the turn of a switch.

Electricity is part of the natural world. The first people only knew it in the form of lightning, a terrifying force that could destroy trees and buildings. The ancient Greeks noticed that rubbing amber, the fossilized gum from trees, made it attract light objects such as feathers or pieces of straw by a force we now know as "static electricity."

For centuries, both lightning and static electricity remained mysteries. No one suspected that they were examples of a form of energy that could be harnessed and used. But in describing the effects of rubbing amber, the Greek scientist Thales of Miletus (624–565 B.C.) used the Greek word for amber, "elektron." The mysterious force now had a name.

THE ELECTRICITY MAKERS
After about 1600, the study of science began to develop in Europe, and many scientists experimented with electricity. It was found that a machine in which a

△ *Benjamin Franklin, the great American scientist, proved that lightning is a form of electricity, by flying a kite in a thunderstorm.*

THE LIGHTNING ROD

A flash of lightning can contain over a million volts of electrical energy producing temperatures as high as 54,000 degrees Fahrenheit (30,000 degrees Celsius). This enormous energy is seeking the most direct way to the ground. When it does so, it can do great damage to trees and buildings and cause fires.

The lightning rod, invented by Benjamin Franklin, is based on the simple idea of providing lightning with a direct route to the ground. The rod is a copper pole fixed to the highest point of a building. A strong cable leads down from the pole and is buried deep in the ground. If lightning strikes, it hits the highest point on the building, the lightning rod. The energy travels down the wire to the ground harmlessly, while the building is unaffected.

piece of cloth rubbed continuously against a glass plate could produce a flow or "current" of electricity. Pieter van Musschenbroek (1692–1761), the professor of physics at Leyden University in the Netherlands, discovered in 1746 that an electric current produced in this way could be stored for a short time in a jar of water and that the spark from it could give an electric shock. His device became known as the Leyden jar.

The next important piece of understanding about electricity came from the United States. Benjamin Franklin (1706–1790) was an American statesman who had many interests, one of which was science. He was particularly fascinated by electricity. It was Franklin who first proved that lightning was a form of electricity.

During a thunderstorm in 1752, Benjamin Franklin carried out an

Hans Christian Oersted.

ELECTRICAL UNITS

Hans Christian Oersted, Alessandro Volta, Michael Faraday, and other pioneers of electricity are remembered in the names given to units and measurements of electricity. An oersted is a unit by which magnetic force is measured. The strength of an electric current is measured in volts. Electromagnetic units are measured in farads, named after Michael Faraday.

Other scientists whose names are remembered in electrical terms include George Simon Ohm (1789–1854), whose name is given to the unit of electrical resistance, and André Marie Ampère. One ampere, or amp, is the amount of current that one volt can send through a resistance of one ohm.

Strangely enough, the electrical term which is used most often takes the name of James Watt, the steam engine pioneer who died in 1819 just before the age of electricity began. The watt is the unit that measures the rate at which electricity is generated or used.

experiment. He flew a kite into the storm clouds, with a key tied to the end of the string. Lightning hit the kite, and a current of static electricity flowed down the string into the key, and from the key to the ground, causing a series of sparks. Franklin then connected the kite to a Leyden jar, and found that the water in the jar became electrically charged.

Franklin also worked out that there were negative and positive electrical charges, which caused sparks when they were brought together.

Another university professor who became interested in electricity was an Italian, Alessandro Volta (1745–1827). Working at the University of Pavia in northern Italy, he invented a number of devices for storing electricity. One was the "electrophorus," which produced a charge of electricity in a metal plate held above a charged piece of ebonite, a hard substance made from rubber.

THE BATTERY

Volta's next step was to produce the "Voltaic pile," the first chemical battery for storing electricity. It consisted of a number of copper and zinc disks separated by paper soaked in acid. A later version had sheets of copper and zinc placed in an acid solution. The sheets were connected above the solution by wire, and a continuous electric current flowed along it.

Although the development of the convenient battery that we know today was many years ahead, Volta's experiments demonstrated the principle on which they work. As yet, however, no one had any idea of how electricity could be put to use.

▷ *Alessandro Volta experimenting with the Voltaic pile. The diagrams on the left show how his "electrophorus." worked.*

◁ *Michael Faraday holds up a magnet with wire coiled around it. The electrical current generated in the wire was the first electrical motor. On the table are two early generators.*

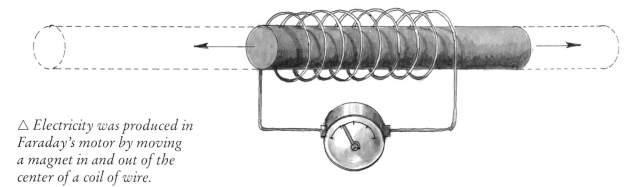

△ *Electricity was produced in Faraday's motor by moving a magnet in and out of the center of a coil of wire.*

MAKING ELECTRICITY WORK

Explanations of electricity were slowly being gathered together. Another contributor to this understanding was a Danish scientist, Hans Christian Oersted (1777–1851). In 1820, he discovered the link between electricity and magnetism. He found that when a compass was placed near a wire carrying an electric current, the compass needle moved. This was a very significant discovery. It showed that electrical energy could be converted into mechanical energy. Electricity could make things move.

A French physicist, André Marie Ampère (1775–1836), followed up Oersted's experiments and made a further discovery. The whole area around the current-carrying wire had a magnetic effect. Ampère called this the "magnetic field." At the time, this seemed like just another interesting fact about electricity. But it was the discovery that led to the development of the electric motor.

The scientist who made the breakthrough was Englishman Michael Faraday (1791–1867). In 1821, he set up an experiment in which an electric current was passed between two beakers of mercury, each containing a bar magnet. When the current was flowing, one of the bar magnets began to rotate round the wire in its beaker. In the other beaker, the magnet was fixed, and the wire rotated round the magnet. When the current was switched off, the movement stopped.

Like all good scientists, Michael Faraday was forever asking questions. His 1821 experiment had used electricity to produce magnetism, which in turn produced movement. Could magnetism, he asked himself, produce electricity? In 1831, he found the answer.

THE FIRST ELECTRIC MOTOR

Faraday made a coil of wire around a magnet, and found that the magnet induced, or brought about, a current in the coil. Almost as soon as it was induced, the current stopped, but if the magnet was removed a current was induced again, this time in the opposite direction. So if the magnet was moved in and out of the coil, a continuous current, running first one way and then the other, was the result. This kind of electricity, called "alternating current," is the kind that is supplied to our houses by power stations. Faraday had made the discovery that made electricity a useful source of power.

As often happens when people in different places are working along the same lines, an American scientist, Joseph Henry (1797–1878), had made the same discovery as Faraday at about the same time. It happened that Faraday was the first to publish details of his experiments, and as a result his name is better known.

INVENTORS TAKE A HAND

Scientists are usually interested in how and why things happen. They are less interested in turning what they have discovered to practical use. This was so with Faraday and Henry. For example, Henry suggested that his work could lead to the electric telegraph, but he did not follow this up himself. Similarly, Faraday did not go on to produce a dynamo or generator to make a continuous supply of electricity. What happened at this point was that various people, almost all of them outside universities, began to use the scientists' discoveries to invent the major electrical devices. In Belgium, Zenobe Theophile Gramme (1826–1901) made the first generator producing "direct current." In the U.S., Samuel Morse (1791–1872) produced the first successful electric telegraph system. Working independently, Britain's Joseph Swan (1828–1914) and the United States' Thomas Alva Edison developed the electric lightbulb.

The new electrical devices worked, and there was a demand for them, but who was going to supply the electricity? Should every home and factory, or every street, have its own

▷ *Edison's lightbulb was lit by passing an electric current through a carbon filament within a vacuum.*

generator? Or could electricity be supplied from some central point, in much the same way as water was piped from the waterworks to every building in the district?

The idea of supplying electricity to a district from a power station seems obvious to us, but in the 1880s it was not as straightforward as it seems today. One problem was the expense of laying miles of copper cable to carry the supply. Another was that most towns and cities already had gas supplies piped beneath their streets which were used for lighting, heating, and cooking in homes. The gas companies did not welcome rivals for their business. Most challenging of all, there were technical problems, especially that a large amount of energy was lost as the electricity traveled along the cables.

MAKING CONNECTIONS

Thomas Edison was not just an inventor. He was also a businessman, keen to make money out of his inventions. Soon after he had produced his first successful electric lightbulb in 1879, he began to plan a district power system. He designed his own cables and circuits, a new generator which was more efficient than previous models, and even a meter to record how much electricity consumers used.

The Edison Electric Light Company built a power station at Pearl Street, in New York's Manhattan district, and wired up offices and buildings within the area. At 3.00 p.m. on September 4, 1882, the electric power was switched on for the first time.

RIVAL SYSTEMS

Revolutionary and exciting though Edison's Pearl Street scheme was, more work was needed before electricity generation was efficient enough to supply a whole town or city. One question that had to be settled was whether direct current (DC) or alternating current (AC) should be used. Edison's was a direct current system. It transmitted a low-voltage current which was particularly prone to the loss of energy.

Joseph Swan.

Another American engineer, George Westinghouse (1846–1914), developed a rival AC system, based on the discovery that high-voltage current lost far less energy in transmission. In the Westinghouse system, electricity was transmitted at a high voltage which was then reduced to a safer, lower voltage before it reached the consumer.

The key device in this system was the "transformer," invented in 1885 by a New York engineer, William Stanley (1858–1916). This increased the voltage of electricity leaving the power station for transmission along the cables. A second transformer decreased the voltage to a safe level for use. The transformer, which works only with alternating current, consists of two coils of wire wound around an iron core. It enables the voltage of the current entering one of the coils to be increased or decreased according to the number of turns in each coil.

For some years, the Edison and Westinghouse systems were rivals. As more electrical devices were invented, including electric trains which first appeared in Germany in 1879, it became clear that both AC and DC had their

▽ *Edison's "invention factory" at his home at Menlo Park.*

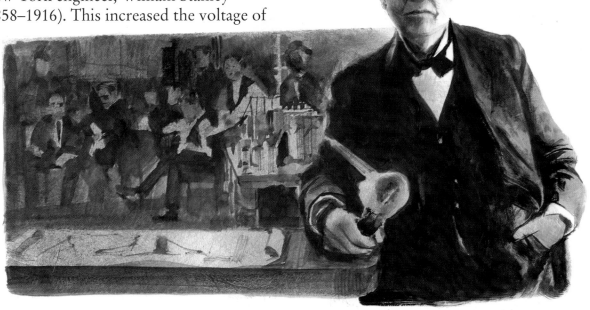

AC/DC
The difference between direct current and alternating current is easy to remember.

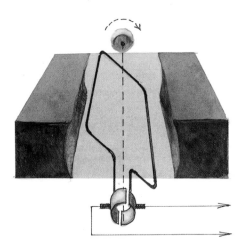

Direct current is the kind produced by a battery. The electricity flows in only one direction, along the positive wire and back through the negative.

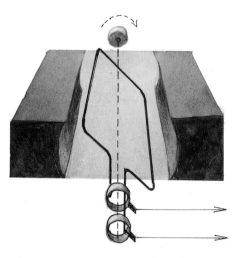

Alternating current, produced by a dynamo or generator, is the kind delivered by public electricity supply systems. The direction in which alternating current flows changes 120 times every second, sixty times in one direction and sixty times in the other.

particular uses. Edison lost interest in electricity and moved on to other inventions. Meanwhile, one of his chief engineers, Croatian-born Nicola Tesla (1856–1943), quarreled with him and left to join Westinghouse. Tesla was an enthusiast for AC systems, and the partnership between him and Westinghouse shaped the future of the electricity supply industry.

In 1893, the Westinghouse company provided the electricity supply for the World's Fair in Chicago with a system that could supply both AC and DC, at different voltages for different purposes.

HYDROELECTRICITY
Two years later, there was another triumph for Westinghouse. In 1885, the world's first hydroelectric power station, generating electricity from the energy of falling water, had been opened in France, and other stations in Europe followed. George Westinghouse's plan was to harness the power of one of the world's most famous waterfalls, Niagara, on the border of the United States and Canada.

The scheme was a huge success. The Niagara Falls power station produced enough energy for local needs and supplied the industrial town of Buffalo, twenty miles (thirty-two kilometers) away. Hydroelectricity had proved itself, and became a major source of energy in countries where heavy rainfall and mountainous terrain make it possible.

THE STEAM TURBINE
Elsewhere, electricity had to be made by raising steam to drive the generator. The first generators were heavy, noisy, and vibrated so badly that the buildings housing them had to be built massively. In 1884, a British engineer, Charles Parsons (1854–1931), demonstrated the

A hydroelectric power station harnesses the force of a flowing waterfall and turns it into electricity.

lighter and more efficient steam turbine which was to become standard generating equipment. The steam turbine contains sets of thin blades mounted on a shaft. Steam is forced at high pressure through the blades, making the shaft rotate. The steam is raised in a boiler by burning coal, oil, or gas, or by using nuclear fuel.

ELECTRICITY FOR EVERYONE

By 1900, everything was in place for the development of the electricity supply industry of today. William Stanley had gone on to work on systems for transmitting electricity over greater and greater distances. A German-born

American, Charles Proteus Steinmetz (1865–1923), also contributed by working out how AC circuits behave and by inventing a device which limited lightning damage to overhead transmission lines.

The erection of transmission lines across the country enabled electricity to be generated at power stations close to their source of fuel, which in turn cut the price of energy by doing away with fuel transport costs. Every industrialized country now has its chain of power stations connected with consumers through a network of high-voltage cables, bringing light and power even to remote communities.

NUCLEAR POWER

A scientific quest to find the smallest particle of matter led to the discovery of an energy source with the ability to supply a large part of the world's energy needs, or to destroy the world altogether.

Until less than a hundred years ago, scientists believed that the atom was the smallest particle of matter and it was the basic building block from which everything was made. This idea had been put forward about 400 B.C. in ancient Greece, and had remained unchallenged ever since. But in 1897, a British scientist, Joseph John Thomson (1856–1940), reported that in his experiments he had observed smaller units of matter that he called "electrons." This led to the realization that each atom was made up of a number of smaller, or "subatomic," particles.

POSITIVE AND NEGATIVE

At that time two different types of particles were identified, "protons" and "electrons," but a third type, "neutrons," was added later. Protons are particles carrying a positive electrical charge, and they are found in the "nucleus" or core of the atom. Orbiting round the nucleus are negatively charged particles called

△ *Rutherford's apparatus for splitting nitrogen atoms.*

1896 Antoine Becquerel (1852–1908) discovers that uranium emits invisible radiation.

1900 Max Planck (1858–1947) comes up with "quantum theory," that energy exists in small, exact units.

1913 Neils Bohr (1885–1962) describes the atom as a nucleus with orbiting electrons.

1919 Ernest Rutherford (1871–1937) deduces the presence of neutrons in the atom's nucleus.

electrons. Neutrons, which carry no charge at all, were not identified until 1932 by James Chadwick (1891–1974), although Ernest Rutherford had earlier said that they existed in the nucleus of the atom. These discoveries meant that theoretically the nucleus of an atom could be divided into two or more parts by a process called "fission."

This discovery completely overturned the physicists' view of the world they thought they had understood, but another shock was yet to come. In 1906, the Swiss physicist Albert Einstein (1879–1955) published his theory that energy and matter were different forms of the same thing, and that each could be converted to the other and back again.

THE SECRET OF URANIUM-235

So far, this was all theory, but the possibilities that lay behind it were enormous. Just how enormous became clearer when a group of scientists began working with atoms of the element uranium. Uranium atoms occur in three different types, or "isotopes," with varying numbers of neutrons inside them.

In 1938, three physicists began experimenting in Berlin, Germany, with the isotope called uranium-235. They were a German, Otto Hahn (1879–1968), and two Austrians, Lise Meitner

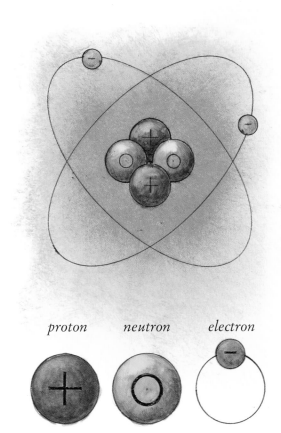

proton   neutron   electron

◁ *The structure of the atom.*

1932 James Chadwick discovers neutrons in the nuclei of atoms.

1939 Otto Hahn produces the first nuclear fission reaction.

1939 Lise Meitner produces energy from nuclear fission.

1942 Enrico Fermi creates the first nuclear reactor in Chicago, Illinois.

(1878–1968) and Otto Frisch (1904–79). They found that if an atom of uranium-235 is bombarded with neutrons it undergoes fission. It breaks down into two equal parts, with two or three neutrons left over. These neutrons act on other uranium atoms, causing them to break up. This effect is called a "chain reaction" and releases a great deal of energy. A chain reaction in which all the nuclei in a piece of uranium explode in a split second results in a devastating explosion.

THE NUCLEAR BOMB

In 1939, World War II broke out, and immediately, both sides stepped up their research with the aim of being the first to produce a bomb which made use of the chain reaction caused by

splitting uranium-235 atoms.

In the United States, Leo Szilard (1898–1964) and Enrico Fermi (1901–1954) managed to create a controlled, continuous source of nuclear energy, a "nuclear reactor," in 1942. In August 1945, the world awoke to the significance of this work when U.S. Air Force planes dropped nuclear bombs on Hiroshima and Nagasaki, in Japan, bringing World War II to an end.

When peace came, scientists turned their attention to the peaceful uses of nuclear energy. The first working nuclear power station was opened at Calder Hall in northern England in 1956. It used the heat produced in a nuclear reaction to

◁ *Nuclear fission.*

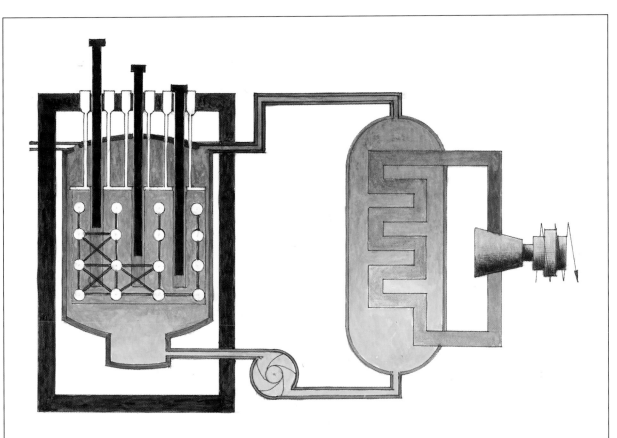

HOW A NUCLEAR REACTOR WORKS

There are several different types of nuclear reactor in use in power stations. The type in the diagram above is called a "thermal reactor." The reactor itself is a large concrete vessel into which rods of uranium are lowered from the top. Tremendous heat is generated when streams of neutrons strike the uranium and split its atoms. Hot gases are channeled to a boiler where they are used to create steam, which then drives a turbine. Meanwhile the cool gas is returned to the reactor to be heated again. Control rods which absorb the stream of neutrons and prevent it from reaching the uranium are used to slow down the reaction or stop it altogether.

create steam which then drove steam turbines to produce electricity.

THE PRICE OF POWER

By 1986, there were over 370 nuclear power stations in the world, with many more being built. But in that year something happened which gave the world a dramatic warning of how easily nuclear power could lead to disaster. One April night, engineers at the Chernobyl power station in the Ukraine ignored safety rules and allowed the station's nuclear reactor to heat up out of control.

The reactor exploded, creating radioactive dust which was carried on the wind and affected human, animal, and plant life across a wide area of northwestern Europe. Thousands of farm animals had to be destroyed, while, closer to Chernobyl, the number of people who will die young as a result of the radioactive fallout may never be known. It was a grim reminder that although nuclear reactors can provide the world with energy that it desperately needs, the price of this energy must be total safety.

PLASTICS

Today, plastics are among the most useful and versatile materials we have. They are used everywhere, from packaging and fabrics, to machine parts and even artificial body parts, but 150 years ago, they were unknown.

A rtificial materials such as plastics are not natural products like wood or cork. They are made by chemical processes. In the past fifty years, plastics have taken the place of natural materials for all kinds of uses, from packaging and clothing, to toys and machinery parts.

IMPROVING ON NATURE

The plastics story begins, oddly enough, with a natural material, rubber. Rubber is the sap of a tree that originally grew in the jungles of South America, though most of today's rubber is grown in plantations in Malaysia. Produced by cutting slits in the bark of the rubber tree, it has been known for at least 700 years as a useful material.

Chemists know rubber as a "polymer." This means that it is made up of long, chainlike molecules that can slide easily over each other. This gives rubber its elasticity. It can be stretched but will then return to its original position and size.

△ *Latex is extracted from trees by tappers to make natural rubber.*

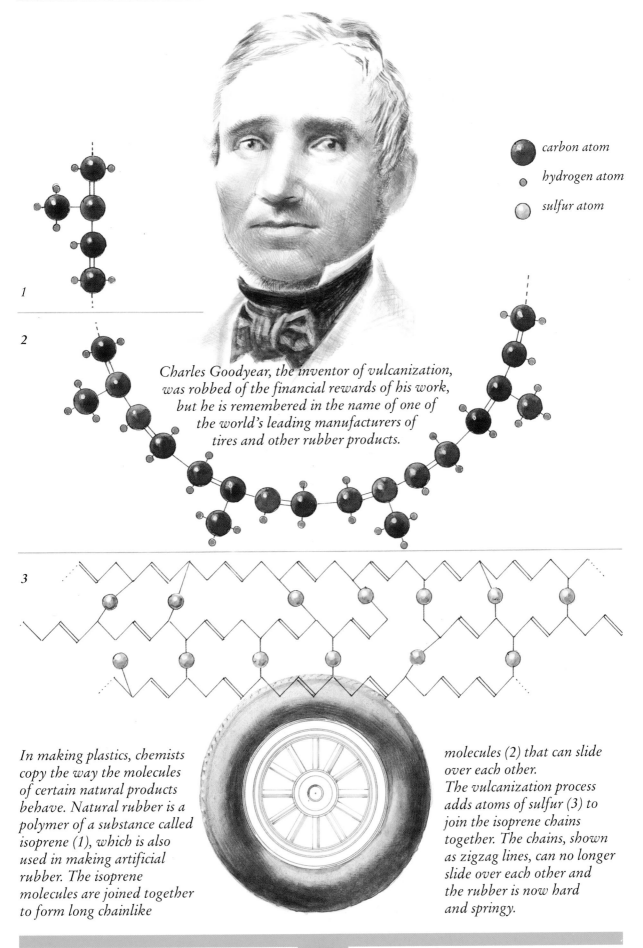

carbon atom

hydrogen atom

sulfur atom

1

2

Charles Goodyear, the inventor of vulcanization, was robbed of the financial rewards of his work, but he is remembered in the name of one of the world's leading manufacturers of tires and other rubber products.

3

In making plastics, chemists copy the way the molecules of certain natural products behave. Natural rubber is a polymer of a substance called isoprene (1), which is also used in making artificial rubber. The isoprene molecules are joined together to form long chainlike

molecules (2) that can slide over each other. The vulcanization process adds atoms of sulfur (3) to join the isoprene chains together. The chains, shown as zigzag lines, can no longer slide over each other and the rubber is now hard and springy.

result was a material that charred but did not melt. Goodyear worked out a way of producing this material in larger quantities, and then set about putting it on the market. He called the process "vulcanization," after Vulcan, the Roman god of fire.

No one in the United States was interested in Goodyear's invention, so he took it to England. He had not patented his idea there, and it was copied by a British manufacturer, Thomas Hancock (1786–1865). It was Hancock who made money from vulcanized rubber, while Goodyear returned to poverty in the United States. Vulcanized rubber became even more popular after 1846, when Alexander Parkes (1813–90) developed a method that was suitable for making thin rubber sheeting for such products as balloons and rubber gloves for doctors.

However, natural rubber has the great disadvantage that it is affected by temperature, becoming sticky when hot and losing its elasticity when it is cold.

American inventor Charles Goodyear (1800-60) became interested in rubber in 1834 and began experimenting with ways of making it less susceptible to heat. Over the next five years, he tried mixing various chemicals with rubber in the hope of finding a mixture that would be stable in all temperatures. In 1839, he hit upon the answer by accident. He had been warming a mixture of rubber, sulfur, and white lead, but let it overheat. The

THE FIRST TRUE PLASTIC

Vulcanized rubber was a manufactured material, but it was based on treating a natural one. Parkes went on to experiment with true plastics, made entirely from chemicals. One of his interests was photography, then in its infancy and using chemically coated glass plates. He made a material called "collodion," a mixture of cellulose nitrate, alcohol, and ether, which was the first photographic film. Another of his inventions was a mixture of cellulose nitrate with camphor, which produced a hard but flexible transparent material. He called this "Parkesine."

The uses of Parkesine were demonstrated at the International Exhibition in London in 1861. Visitors saw Parkesine buttons, combs, pens, decorative boxes, and jewelry. It looked as if Parkes was about to make his fortune, but things went wrong. He was a brilliant chemist, but no businessman, and his company went bust.

REPLACING IVORY

Meanwhile, in the United States, billiard and domino players were getting worried. Billiard balls and dominoes were made of ivory from elephant tusks, which was in short supply and very expensive. A billiard ball manufacturer offered a prize for an artificial material which could be used instead.

The challenge was taken up by a New York printer and part-time inventor, John Wesley Hyatt (1837–1920). His invention was very similar to Parkesine, but he gave it the name "celluloid." Fighting off charges that he had stolen Parkes's invention, Hyatt went into business, making not only billiard balls and dominoes but a wide range of other products. Celluloid collars and cuffs were particularly successful. In those days, the cities were full of soot and ash from coal fires and there were no washing machines, so washing was a difficult task. To look smart, men wore a fresh set of easily washed, detachable celluloid collars and cuffs with their shirts each day.

▷ *J.W. Hyatt with the celluloid billiard balls and dominoes he invented.*

By now, it was clear that there was a market for new artificial materials, and many chemists tried to produce ones that would extend the range of uses. For example, the growth of telegraph and telephone services, and later of electric power transmission, created a need for insulators to carry the lines from pole to pole. Plastics made of nonconducting material were ideal.

THE FILM INDUSTRY

Meanwhile, an American inventor and businessman, George Eastman (1854–1932), created a completely new industry when he adapted Hyatt's celluloid as the first successful roll film for his Kodak cameras. Celluloid film was used at the beginning of the film industry, but its disadvantage was that it was highly inflammable. After a series of horrifying cinema fires, a nonflammable plastic film made of cellulose acetate was invented in 1910.

△ Wallace Carothers

△ The process of spinning rayon out from tree cellulose.

RESISTING HEAT

Another important new material was
invented by Leo Baekeland (1863–1944),
a Belgian-born chemist who emigrated
to the United States in 1889. Like so
many young men with ideas, he saw
America as the place where his talents
would be recognized. He successfully
worked with George Eastman on
photographic materials, which made
him enough money to follow up his
own interests. In 1906, Baekeland began
experiments with mixtures of phenol
and formaldehyde, aiming to produce
an artificial version of "shellac," a hard
coating derived from insects. He called
his product "Bakelite."

Bakelite was something new in
plastics. It could be set permanently by
heating it to a high temperature, after
which it became heat-resistant. It also

resisted water, was an electrical insulator,
and could be easily cut to shape. These
properties made it ideal for a wide range
of uses, such as cups, electrical sockets,
and toilet seats. Soon, there were few
homes which did not contain several
items made of Bakelite.

PLASTICS FROM PETROLEUM

One of the great growth industries
of the twentieth century has been
petrochemicals. In the process of
refining gasoline and other products
from crude oil, a number of by-products
are produced. In the 1920s, chemists
began to be interested in using these by-
products to produce plastics. Some
concentrated on the process called
"polymerization," combining the
molecules of certain chemicals so that
the material produced had some of the

qualities of natural polymers such as rubber. A German professor of chemistry at the University of Freiburg, Hermann Staudinger (1881–1965), first produced polystyrene, now widely used as a building material and in packaging.

Other scientists worked on the production of artificial textiles. The first of these, rayon, was invented in 1884 by a French chemist, Hilaire de Chardonnet (1839–1924). Rayon was made from cellulose, the fiber found in plants, dissolved in chemicals. The mixture was then forced through holes in a machine called a "spinneret," emerging as thin filaments when were then twisted or spun into thread. The same process is used for producing many artificial fibers today.

An American, Wallace Carothers (1896–1937), invented nylon in the 1930s,

POLYMERIZATION
The substance ethylene can be polymerized to make different plastics.

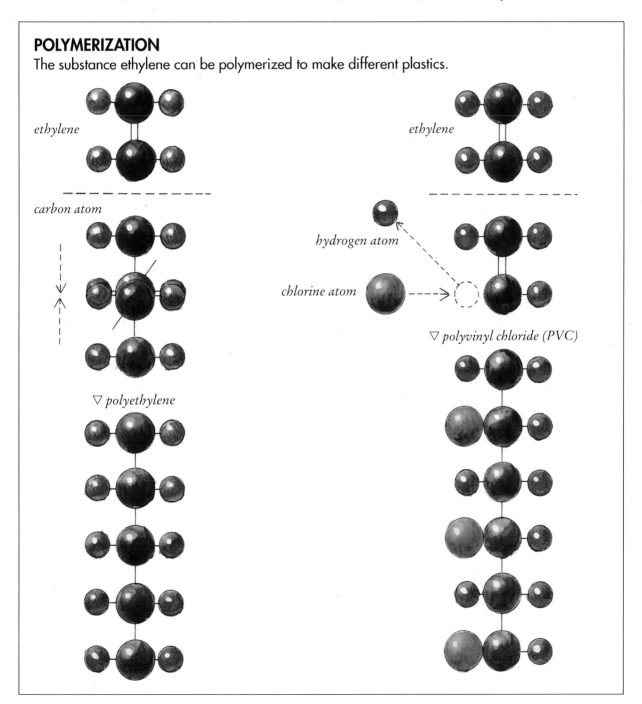

ethylene

carbon atom

△ polyethylene

ethylene

hydrogen atom

chlorine atom

▽ polyvinyl chloride (PVC)

THROW IT AWAY!

One of the great advantages of plastics is that they are durable and do not rot or rust. But this also raises problems when plastics become waste. Many are not biodegradable. This means that they are not affected by weather, the sun's rays, or the atmosphere. Many cannot be burned because they give off toxic fumes. If they are thrown away, they simply become litter.

This has led chemists to a new search, for plastics that, while keeping their important properties of durability, are either biodegradable or can be recycled. One example of a plastic that can be recycled is high density polyethylene, often used for bottles of dishwashing liquid and similar products. This can be used again to make such things as plastic fence posts and trash bags.

using carbon and hydrogen from the hydrocarbons in gasoline, nitrogen from the atmosphere, and oxygen from water. Nylon was the most successful artificial fiber so far. It was the perfect replacement for silk in many products. Silk is so fine that it can be used where many other fibers cannot, but it is very expensive and not very durable. When nylon stockings went on sale for the first time in the United States in 1940, they sold at the rate of a million pairs a day. Nylon is also used for many other purposes, ranging from ropes and parachutes to industrial parts.

Since the 1940s, development work has led to many other artificial fibers. They have various useful properties, such as resisting wear, stains, or creases when woven into materials. Natural and artificial fibers are often spun together to make material. Mixtures of polyester and cotton, for example, are often used for clothes and sheets, combining the feel and appearance of cotton with the durable, crease-free qualities of polyester.

PLASTICS TO ORDER

There are now hundreds of different plastics which have been devised for special purposes. "Neoprene," an artificial rubber invented by Wallace Carothers, is used in vehicle tires. "Thiokol," which resists damage by solvents, is used for flexible fuel lines such as gas pump hoses. "Butyl" is used for tire inner tubes and for lining tanks containing gas or chemicals. The list is almost endless. The science of plastics has now reached the point where a manufacturer needing a new plastic for some special purpose can ask a chemist to tailor-make one.

▷ *Plastics can be cast into any shape, making them useful for the manufacture of a wide variety of products.*

FURTHER READING

Burnie, David. *Machines and How They Work*. New York: Dorling Kindersley, 1991.

Cousins, Margaret. *The Story of Thomas Alva Edison*. New York: Random House, 1981.

Groves, Seli and Dian D. Buchman. *What If? Fifty Discoveries That Changed the World*. New York: Scholastic, 1988.

Gutnik, Martin J. *Electricity: From Faraday to Solar Generators*. New York: Franklin Watts, 1986.

Hawkes, Nigel. *Nuclear Power*. New York: Franklin Watts, 1984.

Higham, Charles. *The Earliest Farmer & the First Cities*. Minneapolis, MN: Lerner, 1977.

Leuzzi, Linda. *Transportation*. New York: Chelsea House, 1995.

Macauley, David. *The Way Things Work*. New York: Dorling Kindersley, 1988.

Michael, Duncan. *How Skyscrapers Are Made*. New York: Facts on File, 1987.

Ostler, Tim. *Skyscrapers*. New York: Franklin Watts, 1988.

Parker, Steve. *The Random House Book of How Things Work*. New York: Random House, 1991.

Payne, Sherry. *Wind and Water Energy*. Milwaukee, WI: Raintree, 1985.

Ritchie, David and Fred Israel. *Health and Medicine*. New York: Chelsea House, 1995.

Smith, Norman F. *Wind Power*. New York: Putnam, 1981.

Weiss, Harvey. *Machines and How They Work*. New York: Harper & Row, 1983.

Whyman, Kathryn. *Plastics*. New York: Franklin Watts, 1988.

INDEX

$E=mc^2$